'Brian Clegg is the master of accessible science writing, and in this beautifully broad and comprehensive new book he gets right to the heart of what makes us what we are. Read it!'

Angela Saini, author of *Inferior* and *Superior*

WHAT DO YOU THINK YOU ARE?

The Science of What Makes You *You*

BRIAN CLEGG

Published in the UK in 2020
by Icon Books Ltd, Omnibus Business Centre,
39–41 North Road, London N7 9DP
email: info@iconbooks.com
www.iconbooks.com

Sold in the UK, Europe and Asia
by Faber & Faber Ltd, Bloomsbury House,
74–77 Great Russell Street,
London WC1B 3DA or their agents

Distributed in the UK, Europe and Asia
by Grantham Book Services, Trent Road, Grantham NG31 7XQ

Distributed in the USA
by Publishers Group West,
1700 Fourth Street, Berkeley, CA 94710

Distributed in Australia and New Zealand
by Allen & Unwin Pty Ltd,
PO Box 8500, 83 Alexander Street,
Crows Nest, NSW 2065

Distributed in South Africa
by Jonathan Ball, Office B4, The District,
41 Sir Lowry Road, Woodstock 7925

Distributed in India by Penguin Books India,
7th Floor, Infinity Tower – C, DLF Cyber City,
Gurgaon 122002, Haryana

Distributed in Canada by Publishers Group Canada,
76 Stafford Street, Unit 300
Toronto, Ontario M6J 2S1

ISBN: 978-178578-623-5

Typeset in Fournier MT by Marie Doherty

Printed and bound in Great Britain
by Clays Ltd, Elcograf S.p.A.

For Gillian, Rebecca and Chelsea

ABOUT THE AUTHOR

Brian Clegg is the author of many books, including most recently *Dark Matter & Dark Energy* (2019) and *The Reality Frame* (2018). His *Dice World* and *A Brief History of Infinity* were both longlisted for the Royal Society Prize for Science Books. Brian has written for numerous publications including *The Wall Street Journal*, *Nature*, *BBC Focus*, *Physics World*, *The Times*, *The Observer*, *Good Housekeeping* and *Playboy*. He is the editor of popularscience.co.uk and blogs at brianclegg.blogspot.com.

www.brianclegg.net

CONTENTS

ACKNOWLEDGEMENTS

Thanks, as always, to the team at Icon Books including Duncan Heath, Robert Sharman and Andrew Furlow. I began writing popular science through reading popular science – and particularly with this book I would like to thank the popular science writers who have made wide areas of science outside of my own direct experience fascinating to me. These include early influences, such as Patrick Moore, John Gribbin and Simon Singh and the authors of all the books listed in the further reading section.

1

A COMPLEX WEB

In the introduction to my 2012 book *The Universe Inside You*, I asked the reader to stand in front of a mirror and look at his or her body, using this experience as a starting point for an exploration of wider science.* In *What Do You Think You Are?*, we are going to turn this idea on its head and go far deeper – discovering the scientific basis of what makes you uniquely *you*. What makes you different from other humans, other animals, plants or even rocks. What is it that makes up the definitive combination of factors that is you?

There are huge similarities between humans, but each is a unique organism – you included. So why is this the case? What makes you the way that you are and different from everyone else? These are questions that we can explore on a whole range of levels. It is easy but unrewarding to state that you are unique in some hand-waving fashion. For a clearer understanding we need to employ the tools of science. In his book *The Scientific Attitude*, Lee McIntyre discusses what distinguishes science from non-science or pseudoscience. He believes that it is 'the scientific attitude', made up of two simple components: empirical evidence (based

* As the *Universe Inside You* website www.universeinsideyou.com features some fascinating experiments that reflect a number of the aspects of what makes you a human, we will be making use of them here too – but the two books cover very different ground.

on experiment or observation, rather than on theory or logic), and being prepared to change theories in the face of evidence that conflicts with them. To understand what makes you *you*, we need to employ such a scientific attitude.

Some would say that science is an unnecessary complication, because what make you the person that you are is your soul. Although in a number of countries the majority now have no religious belief, across the world well over half of the population are followers of one religion or another: religions that almost all say that there is more to a human being than can be explained by physical factors alone. Those holding such beliefs may refer to a soul, or a life force or a vital spark – asserting that there is something more to the makeup of an individual human than physics and chemistry, an essential 'something' that many believe transcends death.

There is no scientific explanation for this extra something – but for the majority of believers, the concept of a soul or its equivalent goes beyond the physical: it is supernatural. As such, by definition the soul cannot be explored by science, as science is the study of nature. If you feel that ignoring the possibility of a soul limits our ability to truly explore all that makes you *you*, that's fine. There's nothing in this book that actively counters the existence of a soul. But we can still make a fascinating journey into your individual existence based on what science is able to tell us about humans, where they came from and how they function.

At the most basic physical level, you are composed of atoms. Everything about your body, from the structure of your cells to the intricate operations of your brain, involves the interaction of atoms in both simple and complex molecules, providing a vast and intricate dance of cause and effect that comes together in the emergent principle we call life.

We perhaps should spend a moment on that 'e' word – emergent – because it is a very important concept, not only when thinking about life, but also when considering other aspects of you, such as consciousness. Something is emergent if it comes into being as a result of the collective interactions of components, but isn't present in the individual components. Very few of us would consider that the atoms that make you up are alive – yet collectively, the whole person certainly is.

Life, then, is more than a collection of atoms, which would still be the same atoms if you were minced up as fine as you like and put in a large jar (try not to think about that image too closely). But clearly you could not be the organism you are were it not for the right atoms being available to make you up. Each of the estimated 7,000,000,000,000,000,000,000,000,000 atoms in your body has to have come from somewhere.* And it will turn out that to reach you, each of those atoms has endured a remarkable journey through time and space.

In one sense, taking the atomic view of 'you' we have to admit that you aren't unique. There may be vast numbers of atoms in your body, in a unique configuration, but each atom of any particular chemical element is identical to every other such atom,† and the human body only contains a few dozen different elements. The fact remains, though, that your particular set of atoms is specific to you, each with its own fascinating backstory, were we able to trace that atom through a history that stretches across billions of years. Your exact mix of atoms will have many similarities to those of other humans, but still differs from everyone else's.

* As we'll be dealing with several big numbers, from now on we'll use scientific notation, where the number 7,000,000,000,000,000,000,000,000,000 would be written as the more compact 7×10^{27}. Here $\times 10^{27}$ means 'multiplied by 1 followed by 27 zeros'.

† To be precise, each atom of any particular isotope is identical. Isotopes are simply variants of the same element with different numbers of neutrons in the nucleus. The name, meaning 'same place' was introduced by English chemist Frederick Soddy in 1913, which he explained was 'because they occupy the same place in the periodic table'.

Even the most reductionist scientist has to admit that a human being is more than a collection of atoms. You are alive. And all the evidence is that it was surprisingly soon after the Earth formed that life began. We think there has been life for around 90 per cent of the Earth's 4.5-billion-year existence. How was it possible to go from an accumulation of dust and gases to the basics of life? For that matter, what is life? We wouldn't be able to ask these questions without ourselves being alive, which is a state that appears to universally need water and energy – so we also need to explore where these essentials come from to help make you *you*.

The very earliest life forms were single-celled organisms like bacteria – yet we are far more than such a single cell, however varied bacteria may have become. The next step in discovering what you are is to trace the path from the earliest life to human existence, putting to rest along the way the idea of the 'missing link' between humans and our biological predecessors. Considering your evolutionary past this way inevitably brings in genetics. At first glance this seems to cut down on your uniqueness. You are somewhere between 99 and 99.9 per cent genetically identical to other humans. For that matter, you share about 96 per cent of your genes with a chimp and 60 per cent with a banana.

However, we need to be wary of allowing a reductionist genetics-based approach. Although, as we will discover, genes do have a very significant impact on what makes you the way you are, the comparison underestimates the differences other contributory factors make. You may have a high degree of genetic overlap with chimpanzees, yet there is no doubt that you are distinctly different from the other great apes. As we will discover, you might get a hint in the fact that you differ considerably more in your overall package of DNA, of which genes only form a tiny part.

We know that our species, *Homo sapiens*, has been around for over 200,000 years. Yet very recently on this kind of timescale, we have begun to have a huge impact on the world around us and have transformed the way that we live. Until a few thousand years ago, what made you *you* would have been almost entirely about biology: now it has to take in the constructed and technological world around you too.

And there's more of you to be explored. Because there are intangible but essential aspects to what you are – your consciousness, personality and behaviours. At some point in our evolutionary history, humans gained consciousness, but exactly what this is and how it works is one of the greatest remaining mysteries of science. We all know (or at least we believe) that we are conscious, but pinning down what it is to say that you are conscious and how consciousness works scientifically is a huge challenge. Yet without consciousness, it's hard to see that 'you' exist as an entity at all.

Personality and behaviour too are very significant factors. Anyone who has had a friend or relative who has suffered from a condition such as dementia where personality and behaviour are altered knows just how hard it is to cope with this change. These are fundamental aspects of what makes you *you*. For a long time, there has been an argument over the relative importance of nature and nurture in contributing to your individuality: how much these aspects of you are down to genetics and how much to upbringing. Now, as we shall discover, there is quantitative data that makes it clearer just how this inner 'you' was constructed.

COMING FULL CIRCLE

It might seem reasonable that we begin our exploration with those most basic components of you, the atoms in your body. Instead, though, we're going to start with a very different, much more human approach. Throughout much of history, a person's definitive position in the world was not drawn from molecular biology, psychology or physics, but out of the spiderweb diagram of a family tree. It was this that made the difference between royalty and commoner, landowner and peasant. What made you *you* was explored through genealogy.*

As we will discover, the family tree has its limits – yet it still has plenty of popular power. Genealogy websites flourish, while there's nothing TV likes better than showing us celebrities making a journey into a small branch of their family tree to discover where they came from. Genealogy is the ideal way to start, as it will eventually enable us to come full circle by exploring the true inner aspects of inheritance when we later return to personality.

Famously, on a popular UK genealogy TV show, working-class actor Danny Dyer, who has specialised in playing unsophisticated cockney geezers, discovered with understandable pride that he was descended from royalty. Even though many of us indulge in a little personal genealogy, few can bring into play the resources available to a TV research team and delve back to make a similar discovery. However, there is no need to feel left out.

I can say with absolute confidence that you too have royal ancestry.

* It's not a true science, but at least it is (almost) an 'ology' as Maureen Lipman would have said in the old BT advert.

2

YOUR ANCESTORS WERE ROYAL

Many of us enjoy genealogy. It enables us to get a feel for our close family and to look back a number of years – but the approach can only do so much. The word 'genealogy' comes from an Ancient Greek word meaning 'tracing of descent'. The implication is that your pedigree* defines who you are. In part, having a list of 'who begat whom' was required to determine which family member would inherit an estate after death, but it also became associated with the idea of a person's worth. It was as if the family you were born into somehow defined what you would become in life, an assumption perpetuated and locked in by rigid social structures, such as class or caste systems.

Taking the UK as an example, while the class system has become significantly more diffuse in the last 100 years, some still hold to a distinction based purely on birth – and never more so than when there are royal connections. Traditionally the British divided themselves into working class, middle class and upper class (with some gradation, such as 'upper

* Pedigree is a much more fun word than 'genealogy'. The term comes from the French *pé de grue*, meaning crane's foot. This probably arose from the use of three curved lines in family trees to denote succession, which looked a little like a bird's claws on the page.

working' or 'lower middle').* The borderline between working class and middle class has become extremely diffuse. For example, my father's parents were mill workers – undoubtedly working class. My father didn't go to university and started work in his teens, so also started off working class. However, he took night classes and became a manager and finally a director of the company where he worked his entire career – making the transition to middle class.

The working-class label remains one that is held with pride. However, the boundary is fuzzy, as the majority of 'middle-class' people are no longer in the traditional middle-class professions such as clergy or doctors; nor are they business owners, but typically are employees of an organisation, as much as anyone who regards themselves as working class. By contrast, the remnants of the upper class still define themselves not by their achievements but as a result of the family they were born into, and this is a class that reaches its pinnacle in royalty. It's for this reason that actor Danny Dyer was so excited to find that he was a descendant of the English king Edward III who lived between 1312 and 1377. Dyer was, of course, related to far more individuals who weren't royal, but the remnants of class status made this relationship seem more interesting. As we shall discover, though, it doesn't make Dyer particularly special. Not special at all, in fact.

EXPONENTIAL DOUBLING

Interesting though a family tree may be – and there is no harm in putting one together as entertainment – it's difficult to look at one for long without realising the limitations of the structure. Go back a few generations and the contents of the tree will become very selective. Whoever constructed it will have chosen only a few of the possible branches to

* See the 'Further reading' section at the back for a link to the classic *That Was the Week that Was* sketch on the three British classes.

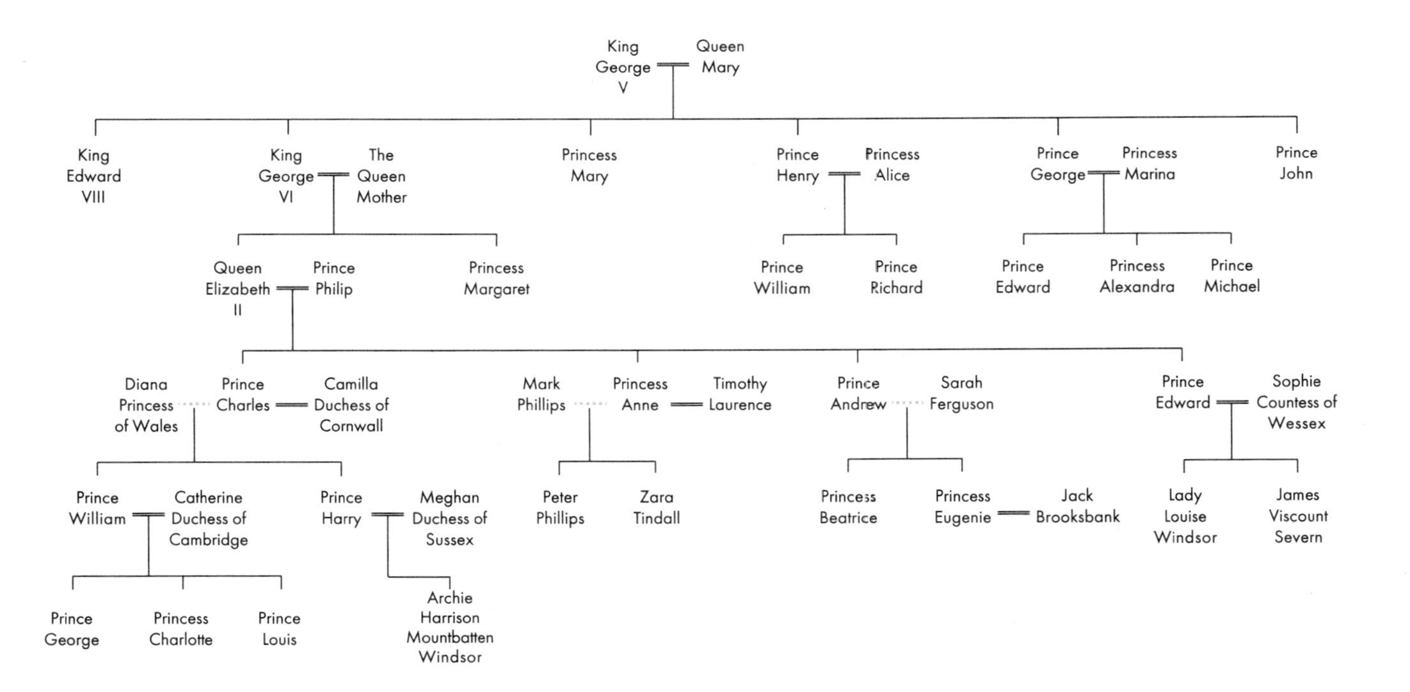

A (royal) family tree – to fit even these few generations many individuals are missing (Source: *Town & Country Magazine*).

pursue. In truth, it's not so much a family tree as a family twig. And there's a good reason for this restriction, arising from the mathematical phenomenon known as exponential doubling.

It's not uncommon these days for 'exponential' to be used to mean extreme – as in 'this is an exponentially large figure' – but in mathematics, exponential has a precise meaning, which is far more dramatic. Normal 'linear' growth involves going up by some multiple of the factor being considered – time that has passed, or generations, or whatever. So, for instance, after five years, something undergoing linear growth might be five times bigger. After ten years, ten times bigger. And so on. But exponential growth is on a different scale. We've already introduced what is known as exponential notation where instead of, say, writing 1,000,000,000 or 1 billion, we use 10^9. Here the number '9' is the *exponent*, the number of times that 10 is multiplied by itself to produce the required value. If each unit of time (or whatever) involves an increase of the exponent, growth is exponential. So, for example, after five years it might be 10^5 times bigger – 100,000 times – while after ten years it might be 10^{10} times bigger – 10 billion times bigger. That's not just getting bigger, but the rate of increase is accelerating dramatically. When something grows in this fashion it rapidly gets out of control.

Rather than raising 10 to the exponent, an alternative type of exponential growth involves exponential *doubling*. Here, the number involved doubles at each step – so after n years (or whatever the factor under study is) the value is 2^n times bigger. Exponential doubling is often illustrated using a story involving a chessboard and grains of rice. According to the legend, as a reward for undertaking a task, a wise man asked a king for an apparently simple payment. All he required was a few grains of rice. The total required a spot of calculation and a chessboard. The idea was to put one grain of rice on the first square of the chessboard, two grains on the second, four grains on the third, eight on the fourth and so on, until all the squares had been loaded up with rice. The total number of rice grains involves exponential doubling.

In the story, the royal dupe who agrees to this deal ends up owing the sage far more rice than exists on the planet. The fact that this isn't obvious reflects our mental inability to cope with the impact of exponential doubling. There are 64 squares on a chessboard, and we start with just a few grains on each square. So, however rational we are, it's hard to get away from thinking that the outcome must be something comparable with 64 times a sizeable but manageable number. Perhaps around 64 million or 64 trillion. The reality, however, is very different.

Let's take a look at the total number of grains. As we have seen, on the first square we put one grain. With two squares there's one grain on the first square and two on the second – three in total. With three squares there are seven grains. And with four squares there are 15. Nothing frightening so far. That sequence of numbers – 1, 3, 7, 15 … – is just one short in each case from a more familiar series: 2, 4, 8, 16 … The powers of the number two. This means we can quickly calculate how many grains there are with n squares occupied this way as 2^n-1. We multiply two by itself n times and take away one. Here we can see very clearly how that exponential part is creeping into the calculation.

So, the total amount of rice required to fill up the whole board would be $2^{64}-1$. Written like that, it still doesn't look too scary, as it's only little old 2 that is being multiplied by itself. But to put it another way, that is 18.5 billion billion (if you want to be precise, it's 18,446,744,073,709, 551,615 grains). A whole lot of rice. Very roughly it's about 300 billion tonnes of the stuff – the amount the world would currently consume in 600 years.

The sheer volume of rice involved is fascinating – but what has this to do with genealogy? Exponential doubling also comes into a family tree due to a simple fact that we're all aware of, even though we tend not to think through the consequences. Each individual person on a family tree – you, for instance – will have exactly two biological parents. This means that if we ignore siblings (we will come back to them) and simply trace back an individual's tree into history, the number of people in each

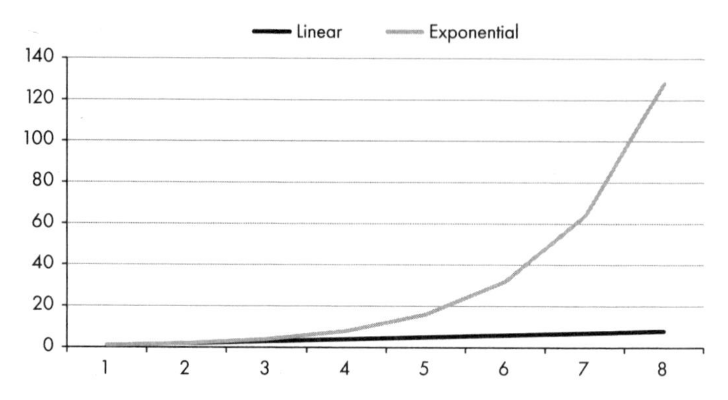

Linear versus exponential growth: the linear line goes up one each time, the exponential line doubles each time.

preceding generation doubles. In your tree you will have two parents, four grandparents, eight great grandparents and so on.*

Just like the chessboard rice grains, these numbers do not initially seem to be very extreme, until we start to combine them with a realistic count of generations going back in time. For convenience, generations used to be treated as 25 years, because it's easy to do the maths and it was a reasonable approximation for the average age at which people became parents. These days, 30 is more like a sensible average, but for most of history the value tended to be lower, so 25 may be the best number. We'll try both.

Just like the chessboard, the count of your directly linked ancestors goes through exponential doubling – in this case doubling with every generation, rather than with every square on the board. There's one of you, three people with your two parents, seven people with your four grandparents and so on – that familiar series of 1, 3, 7, 15… So, looking back n generations in your past, we get to a total of 2^n–1 people, in a tree stretching back into history from you as the root. And this is true for

* You may, of course, not know who all these people were, but it is inevitable that they existed.

every one of the near 8 billion people alive today (which brings in siblings). Each individual has a tree stretching back in history the same way.

THE MISSING BILLIONS

Now for the big reveal. It has been estimated that around 110 billion people have lived since *Homo sapiens* came into existence a couple of hundred thousand years ago. That's obviously a guestimate, but not a bad one. The figure that is most often quoted is 108 billion, based on a calculation by a group called PRB. I've made it 110 billion, partly because 108 billion gives a spurious feeling of accuracy and partly because one of their assumptions, that *Homo sapiens* has been around for 50,000 years, is a significant underestimate based on current data.

So, how many generations does that 110 billion represent? Just using your personal tree, we would need to go back around 37 generations. This is because 2^{36} is around 68 billion and 2^{37} is around 137 billion. If there were a totally separate tree for each person alive today, we would only need about 34 generations. Clearly that's a reduction too far, as siblings will share the same tree – so it would be realistic to go for, say, 36 generations.

If we use 25-year generations that takes us back just 900 years into the past, or with 30-year generations we get back 1,080 years. Using this simplistic calculation, humans should have only been around for 900 to 1,000 years. In reality, though, we know that history goes back several thousand years further and archaeologically and palaeontologically speaking we can say that *Homo sapiens* has been in existence as a species for perhaps 200,000 years. Being generous and using the longer figure of 30 years for a generation, that's 6,666 generations. Which would mean 2^{6666} people in the single family tree starting from you, a phenomenally large number. It's approximately 4×10^{2006}. To put that number into context, the number of atoms in the universe is often estimated to be around 10^{80}.

Clearly, something has gone horribly wrong with this calculation. What these numbers reflect is that the nice well-ordered family tree we get from genealogists has been pruned incredibly tightly. We don't have neat, spreading trees, but complex tangles. Go back a few generations and you will find that branches entwine and interlink in a more and more complex fashion. Increasingly large numbers of the inhabitants of each generation will be duplicated over and over again as the same historical person appears in different branches. And this effect will become stronger as you go further back in time because of low mobility. You don't have to go back very far in history to get to a stage where the majority of people never ventured far from their home village. They would not have the whole world available to them as breeding partners, but rather a tiny gene pool.

Something had to give way to go from more than 4×10^{2006} to a mere 110 billion (for a clearer comparison, that's 1.1×10^{11}). And that's where we can all share Danny Dyer's excitement. It's not that there is anything wrong with his genealogy, but rather that we can say with certainty that *everyone* has royal ancestry, thanks to the application of statistical analysis to the numbers of branches and linkages in our ancestry. If we go back far enough in any such intertwined tree we end up with regional common ancestors – people who for those of a certain region we can guarantee will be in their family tree. This is true of not just a few of the potential ancestors, but vast numbers of them. Go back far enough, in fact, and we can say that you are related to every single person from your region who has living descendants. That will apply to kings and queens, just as much as minstrels and servants, murderers and vagabonds. They're all there in your family tree.

EVOLVING A THEORY

Before we manage to resolve those vast numbers of missing ancestors, we need to get a feel for how far back we need to go. In Chapter 6 we will be looking at where humans came from, but here it's just a case of knowing where to stop looking into the past with those family trees. And to find the right point to make the break, we are going to have to deal with the E word: evolution.

I can't help but feel rather sorry for scientists whose work involves evolution, because there is surely no other scientific topic that has become so charged with emotion.* Physicists might grumble about those who don't share their interpretation of quantum mechanics, or colleagues who spend their careers working on theory that has no resemblance to the real universe – but they are unlikely to be dragged into debates with school boards and legal battles over the reality of their theories. The odd thing from the outsider's viewpoint is how straightforward and obvious evolution is.

We only have to make two assumptions, which are nothing but common sense, and evolution is inevitable. The first is that we have the ability to pass on various characteristics to our offspring, who are not carbon copies of us, because a mix of those characteristics comes from both parents. The second is that organisms with characteristics that help them survive and thrive are more likely to have offspring to whom they can pass on those characteristics. Combine these and you've pretty much got evolution happening whether you like it or not. Darwin didn't know *how* it worked – he didn't know about the genetics we'll meet later on in the book – but it's hard to see how anything else could happen in the biological world.

Science tells us that evolution has led over the billions of years that there has been life on Earth to the proliferation of species. It's quite common for people to say 'I accept micro-evolution – that's obviously going to happen. So, for example, if birds with bigger beaks are better at breaking up the nuts they eat, then over time more of the birds with bigger beaks will be more likely to breed and big beaks will dominate.

* With the possible exception of climate change.

But I don't see how a mouse can turn into a chimpanzee, or a chimp into a human.'

There are two problems here. One is the failure to recognise that 'species' is an arbitrary concept. Every single organism is the same species as its parents.* Which seems to imply that you never will get a new species emerging. However, we're not dealing with a single change, but rather an accumulation of tiny changes to genetic makeup that eventually result in an organism that is not the same species as its earlier ancestors. Exactly how we define a species is a little vague, but the traditional point at which a species divides off from another is when it's no longer possible to interbreed and produce fertile offspring. Depending on the rapidity with which an organism reproduces, such a change could take millions of years or just decades.

A good parallel with the paradox of changing species over time despite an organism always being the same species as its parent is in the colours of the rainbow. We know that there are far more colours than Newton's original, arbitrary seven. Zoom in to the detail of a rainbow and there are millions of subtly different colours – my computer can display a palette of over 16 million. Look at two adjacent colours and they will apparently be identical. (Try this in a paint program on your computer if you don't believe it.) Every one of those 16 million-plus colours looks the same as its 'parent' colour next to it. Yet across the whole spectrum we go from red to orange to yellow to green and so on, with all the variations in between. This is what happens with species too.

The other problem with the argument 'I don't see how a mouse can turn into a chimpanzee, or a chimp into a human' is that you don't need to, because this does not happen. We are not descended from our cousins, the other great apes. Rather, go back far enough and you will reach a common ancestor of both us and our closest living relatives, the

* In principle a radical mutation could result in a member of a new species being born, but the chances of such a major mutation being survivable and able to be passed on to future generations is very small.

chimpanzees and bonobos. Go back further and you will reach a common ancestor that also takes in other great apes. Further still you will find a common ancestor we have with monkeys as well ... and so on, eventually reaching a common ancestor with a mouse. And so on again. It's like an upside-down, back-to-front version of a family tree.

To take in the whole of human genealogy we need to go back to the point in time our species evolved from its predecessor. You could, of course, carry on further and further into those common ancestors – but it's hard enough getting our head around just our human family trees. So, we will sensibly make the break when *Homo sapiens* came into being, around 200,000 years ago. This rough date comes primarily from fossil evidence. We'll come back to our more distant ancestors in Chapter 6.

UNCOVERING MITOCHONDRIAL EVE

So, how do we demonstrate everyone's parallels with Dyer's royal blood? We need a way to look into the distant past, discovering how far back we need to go before we see shared ancestors for large groups of people. One way to do this is to use DNA. We will come back to DNA in a lot more detail in Chapter 9, but for the moment the important thing is that the bulk of your DNA molecules, which make a very significant contribution biologically to what you are, come from both your parents. However, a small amount of your DNA – so-called mitochondrial DNA – comes only from your mother. This is the tiny relic of DNA still remaining in mitochondria, essential parts of your cells which developed from bacteria. Mitochondria are often called the power units of our cells, because they are responsible for producing the molecules that store tiny amounts of energy to be transported around the body.

As the distant ancestors of mitochondria were independent entities, they had their own DNA, distinct from the main DNA of the cells they reside in. Like our chromosomes, each of which comprises a single long molecule of DNA, the DNA in mitochondria contains genes. Over the many millions of years that mitochondria have been in action in humans (and almost all other organisms with complex cells), the genes from the mitochondria have largely been transferred out to our chromosomes. In the case of humans, just 37 genes have been left. But this tiny fragment of DNA is special as it is inherited only from our mothers.

By combining data on variants in the mitochondrial DNA of a range of individuals with the rate at which this DNA was assumed to have mutated,* it was possible to work backwards to deduce when the most recent common ancestor of all current humans was alive. Someone from whom every living person is descended. This so-called 'mitochondrial Eve' is thought to have lived 150,000 years ago, give or take a few ten thousand years. It should be stressed that mitochondrial Eve was neither the only woman living at the time nor the first woman – this process merely identifies a likely distance in time to a female individual who according to mitochondrial DNA was the most recent woman to be in the family tree of everyone now alive.

SEARCHING FOR COMMON ANCESTORS

Finding a timescale for mitochondrial Eve was an interesting exercise, but it doesn't give a good picture of 'what makes you *you*', both because it was based on crude assumptions (some of which are outlined below) and because it only tracked through the female line. To get a better picture of your ancestry, we need both male and female lines – and some impressive statistics. Back in the late 1990s, Joseph T. Chang, Professor of Statistics and Data Science at Yale University, wrote a paper entitled 'Recent Common Ancestors of All Present-Day Individuals'. In it,

* A mutation is a change in the code stored in the DNA, which can be caused by an error in the mechanism used to copy the DNA or by external intervention, such as being zapped by a cosmic ray. All of us have mutated DNA – we aren't talking about something out of X-Men here.

Chang took a trip back in time using a mathematical model that gives a fascinating picture of our heritage as individuals.

Initially, Chang kept some of the extreme simplification of the earlier exercise. The mitochondrial Eve calculation had assumed each generation dies off producing the new one – there is no overlap between generations – and that each individual only has a single parent. It also assumed that an individual's parent is randomly selected from the entire population of the previous generation, clearly a huge over-simplification. (Just think how your parents met – would they have had an equal chance of getting together with every other person of the opposite sex of any age alive at that time? I doubt it.) Chang still simplified reality, but made significant improvements.

Chang was able to demonstrate that once two parents were considered, even using simplifications that make complex tangled families* impossible, the inclusion of the second parent meant that common ancestors cropped up far more recently than was suggested by the female-only line. As Chang puts it 'mixing occurs extremely rapidly in the two-parent model, so that [common ancestors] may be found within a number of generations that depends logarithmically on the population size.'

Let's unpack that word 'logarithmically'. This means that the population size would be around 2^N where N is the number of generations to the most recent common ancestor. For example, in a population of 1 million, the number of generations required to reach a most recent common ancestor is just 20, as $2^{20} = 1{,}048{,}576$. For the same size population, mitochondrial Eve with her single-parent lineage would require thousands of generations.

* As an example of a tangle, my grandfather's stepmother, who had children with my great grandfather was also my grandfather's cousin (give or take a 'removed'). For that matter, the older models used wouldn't allow for the possibility that a man or woman could have children with more than one partner.

This is striking enough, but there is a more remarkable result still from taking this approach. This calculation might give an approximate value for the number of generations to a most recent common ancestor. But this certainly isn't the *only* common ancestor. There may be more than one in that same generation. And for certain there had to be at least two in the previous generation, four in the generation before that and so on – because each of the most recent common ancestors' parents, grandparents etc. would also be our common ancestors.

This back-in-time spread of common ancestors increases exponentially as we get earlier in time. Eventually there must come a point where *everyone* in a generation whose line didn't die out – everyone who has descendants living now – is a common ancestor of today's entire population. At this point anyone in the population we're looking at can say that *everyone* in that ancient generation who has living descendants is their ancestor. Surprisingly it doesn't take long to get back to a generation where this is the case – only around 1.77 times the number of generations required to reach the most recent common ancestor.

Of course, in the real world we have a population of a lot more than a million – at the time of writing about 7.7 billion. If we just plug the current population into the simple formula, that's around 33 generations or 1,000 years to get a common ancestor of all of humanity. But here one of the principle assumptions of the model breaks down – even now, we are significantly more likely to have children with someone born in the same country than from another country, and going back in time this was far more likely still. But the model works on the assumption that anyone in the world can be your parent.

Given those provisos, Chang proved mathematically that his estimates for the number of generations to the most recent common ancestor, and for all the population with surviving descendants to be common ancestors, were realistic. This, however, was only the start. By 2004 he was able to publish in *Nature* a paper with Douglas Rohde and Steve Olson that took in the isolation of populations and the tendency to mate within social groups, providing a much more realistic model of our true ancestry.

The models showed that even in populations with significant internal structures, a most recent common ancestor would still be reached relatively quickly (far quicker than working back to mitochondrial Eve). With this more sophisticated model and conservative assumptions about migrations of individuals, Chang and colleagues came up with a worldwide most recent common ancestor date of 1,415 BC and a date when everyone with surviving descendants was your ancestor of 5,353 BC. With a rather more generous rate of migration and mixing, this can be made as recent as around 2,158 BC.

BRINGING IN THE GENES

Since Chang's paper was published, there has been considerably more work done by other researchers, adding genetic information to the statistical data from the model, reinforcing Chang's conclusions that, for example, within Europe you only have to go back 600 years to hit a most recent common ancestor and 1,000 years for everyone in Europe who has living descendants to have been one of your ancestors.

Of course, we have no idea who the most recent common ancestors of Europeans were – and the mathematics allows for many more than one common ancestor in a generation. But when we go back far enough for everyone who has living descendants now to be an ancestor, we hit royal pay dirt. A handful of people can trace back their ancestry to royalty from 1,000 or more years ago. And given that these people are still alive, that means *everyone* of European extraction is descended from those same royal individuals.

The specific example that tends to be picked out is Charlemagne, who was king of the Franks and Holy Roman Emperor, living from 742 to 814. There is good evidence that he has living descendants – so this means that if you are of European extraction you can claim Emperor

Charlemagne for your family tree just as much as Danny Dyer has his royal background. Similarly, the current British royal family are thought to be descended from William the Conqueror, who is sufficiently far back in history that should you have European descent, he is almost certainly your ancestor too.

If you don't think you have European ancestors, find yourself a suitable ruler far enough back in your own history and you will have a certain hit. And without doubt we are all descended, every one of us, from royalty in prehistoric dynasties who may not feature on the historical record, but existed nonetheless. The earliest known Chinese dynasties are said to stretch back to around 2,000 BC, encompassing the lower figure for ancestors of everyone in the world. Egyptian dynasties are said to go back to around 3,000 BC. Sumer was settled around 5,000 BC – so, the chances are that wherever you are from, you can claim a link to the Sumerian royal family.

There's always the proviso of these common ancestors needing to have descendants still alive. Some won't – but a percentage will. In the end, the individual doesn't matter. The point is that in terms of what you think *you* are, there is no doubt that you and I share common ancestors, and we are both descended from royalty. Don't get too full of yourself, though. You are also descended from murderers, vagabonds and thieves. (Actually, come to think of it, all of the above could apply in the case of the royalty too, and almost certainly did. You didn't get to be royal in the early days without a spot of Machiavellian machination.) Your ancestors were also merchants and beggars, philosophers and artists, saints and sinners. Our true family trees are not the spindly little cherry-picked things we usually see – they are vast, intertwining forests that give each and every human being a rich and wonderful heritage.

Before we move on to take a plunge back into your very oldest origins, we should take a moment to dismiss a genealogical myth that should be obvious from what we've just read about worldwide common ancestors. Biologically speaking, there's no such thing as race.

THE RACE CARD INDEX

We love to split things up – including people – into groups and categories. It's how we understand the world, and often it is very useful. But with people, all too often the categories reflect those who belong and those who don't. We've got plenty of ways of categorising humans that have some validity and usefulness. Although there is no biological basis for nationality, for example, it has legal standing and as such is part of what makes you what you are. Likewise, socioeconomic groupings and culture are not biologically based,* but they certainly exist and are sometimes used to discriminate against individuals and groups. Then there are biological differences, whether at the gross level of gender, or the more sophisticated, but still biological, aspects of being, say, straight or LGBTQ+. We need to recognise all these groupings, both because they are part of our identity and because many of them can be used as a division for discrimination, which a civilised society needs to avoid.

However, race is a different kind of categorisation. There is no scientific basis for the concept of race. It makes neither biological nor cultural sense. Whether at the crude level of black or white, or using groupings such as African Caribbean, European, Southern or East Asian – or, for that matter, old pseudoscientific terms such as Caucasian and Mongoloid – race is an arbitrary notion which lends itself to use by racists. Let's say it one more time. There is no biological concept of race.

Some may think that making this claim is political correctness – that I am ignoring obvious racial differences because I'm a wishy-washy liberal. But science has a clear message. There is far more genetic difference within a racial group of your choice than there is between that imagined group and another one. A major study of 1,056 individuals from 52 populations in 2002 found that 93 to 95 per cent of genetic variation was

* I am grouping religion (or lack of it) in with culture here, as the two are usually very tightly intertwined.

within a population. Yes, there are small genetic differences that apply to regions of origin – but they are far, far smaller than any differences that we just ignore by using race labels.

What then, for example, about the very obvious difference between black and white skin? Of course, some people have darker skins than others, but having a difference in pigmentation doesn't make you part of a separate biological entity. Everyone – absolutely everyone – is a mutant. Each of us is genetically different from the rest of humanity. Even identical twins, who are genetic clones, making their embryos nominally genetically identical, develop biological differences from the word go. So, the mere fact that some people have darker skin pigmentation is nothing to get excited about. When I was younger, my hair was bright red. I had a distinct pigmentation difference (caused by a mutation) from the majority of humanity.* But no one suggested I was a different race from others around me because of this.

Every other 'racial difference' is also biologically trivial. Apart from small visual differences, the majority of our associations with race are based on culture and on socioeconomic factors – but race is far too crude a label to be useful culturally, and irrelevant from a socioeconomic viewpoint. The 'race' label is just an excuse for xenophobia. Remember what we've already seen – you only have to go back a few tens of generations and everyone with surviving descendants is a common ancestor. Turn this on its head and it becomes clear that you are not just related to all those old royals – you are also related to everyone else alive on Earth, whatever race you label them as.

It's entirely natural to be suspicious of groups who are even trivially different from the group we identify as 'us'. Historically, this was a good survival trait. But not everything that is natural is good. It's entirely natural for most children to die before they reach adulthood. As a species we

* The colour of hair and the colour of the skin are caused by variants of the same pigment, melanin.

are only part way through the process of growing up. We all still have a reflexive uncertainty of the 'other'. But keeping the imaginary label of race is not helping with our development.

We've seen, then, that the family tree only gives a very limited view of what makes you *you*. It gives some hints of one aspect of your defining features, your genetic background, but it's probably the least impressive way to do so. So, let's take a totally different viewpoint and travel much further back in time to find how the most fundamental components that make you up came into being. We are all collections of atoms. But what are they, where did they come from, and how did they get into you?

3

STARDUST MEMORIES

We have established that you only have to go back a few tens of generations to establish where you came from in a genealogical sense. But we've much further to go – billions of years – to see where you came from physically. At the basic physical level, you are made up of just four types of particle, with the help of a few forces: the specific particles in your body have been around for the majority of the lifetime of the universe. In one sense this feels very reductionist. You might rightly argue that you are far more than a collection of particles: through emergence and interaction, the whole can, indeed, be far more than the sum of its parts. However, it would be silly to deny that those parts exist and that they are fundamentally the objects that make you up.

BUILDING BLOCKS OF EVERYTHING

So, in tracking down what makes you *you*, it's essential to get a better feeling for those components. The particles in question aren't quite the ones that you may have come across if you only studied science to high school level (or some time ago). The familiar one will be the electron. Like the other three, this is what's described as a 'fundamental' particle, meaning that, to the best of our knowledge, these particles don't have subcomponents. They aren't made of anything simpler. Flows of electrons are what's usually involved when there's an electrical current, and it's the quantity and distribution of electrons around the outside of an atom that determine how it will behave chemically.

Electrons are small. Really small. Some would say impossibly small. Their mass is around 9.1×10^{-31} kilograms.* It takes around a million, trillion, trillion of them to make up a kilogram. Ask a physicist how big an electron is and, if the physicist is honest, he or she might grimace. The official answer is that an electron is a point particle, meaning it has no size at all. If this is true, it causes some serious problems for theory, as some values grow as the radius of an object gets smaller, giving infinite values for an object with zero diameter. (Think of the density, for example, of an object with mass but no volume.) Equally, though, if an electron really did have a very small but measurable radius, that too throws up problems for different aspects of theory. Either way, electrons are pretty weird little entities.

Strange though electrons are, we have known about them significantly longer than we have the other components of the atom. By the late 1890s, electrons had been identified as particles, taking over the name that had been devised in 1891 for the charge in a chemical bond. But the names of the other component particles that make you up – quarks and gluons – would not enter the language until the 1960s. You might be wondering what happened to protons and neutrons. These are the familiar particles that make up the nucleus of an atom, starting with a single proton that is all that there is in the hydrogen nucleus and adding extra protons and neutrons to make up the heavier elements. So, for example, there are 79 protons and 118 neutrons in the nucleus of the

* We've already discovered that 10^n means 10 to the power of n, the same as 1 with n zeros after it. Similarly, 10^{-n} means 1 divided by 10 to the power of n. So 10^3, for example, is 1,000 while 10^{-3} is 1/1,000. This makes 9.1×10^{-31} a way to represent 9.1/10,000,000,000,000,000,000,000,000,000,000 – which hopefully makes it clear why we use scientific notation for small numbers as well as big ones.

most common isotope of gold. But neither the proton nor the neutron is a fundamental particle – each has sub-components.

Every proton is made up of two up quarks and one down quark, held together by a flow of gluons, while each neutron contains one up quark and two down quarks, again linked by gluons. The 'up' and 'down' part of the names simply comes from the way their mathematical representation is written out. The 'quark' part is more interesting. Murray Gell-Mann, the physicist who came up with the name, said that the sound (which is intended to be more like 'kwork' than 'kwark') came to him out of the blue. But later, seeing the line in James Joyce's book *Finnegans Wake* 'Three quarks for Muster Mark!', he adopted that spelling of a word that is still, technically, pronounced 'kwork'. Gell-Mann was inspired to do so as quarks happen to come in threes in protons and neutrons.

Like electrons, quarks are very light particles, with masses of 3.9×10^{-30} kg (up) and 8.4×10^{-30} kg (down), while gluons have no mass at all. This might seem to imply those more familiar protons and neutrons should have masses of 16.2×10^{-30} kg and 20.7×10^{-30} kg respectively, but that would give us two problems. First, protons and neutrons are significantly more similar in mass – and secondly it would make the particles around 100 times too light. The reason for this brings in the world's most familiar equation, Einstein's $E = mc^2$, which makes energy (E) equal to mass (m) times the square of the speed of light (c). The more energy in something, the bigger the mass it has – and the majority of the mass of a proton or a neutron comes not from its component particles, but from the energy that holds those particles together.

BRUTE FORCE WITHOUT IGNORANCE

Thinking of particles being held together makes this a useful point to bring in the other part of what makes up your atoms. As well as the four particles, I mentioned 'a few forces'. Specifically, there are two we need to consider. The more familiar is electromagnetism, responsible for the forces between magnets and between electrically charged objects. Electrons have a negative electrical charge and the nucleus of an atom has a positive electrical charge so there is attraction between them, just as there is a pull between the wires of an electric motor, or an attraction between a statically charged balloon and bits of paper.

This attraction between electrons and the nucleus prevents the electrons flying off and doing their own thing (though in good conductors, such as metals, some electrons do manage to escape and do just that). But it might seem at first glance that the relationship between a negatively charged electron and positively charged nucleus would be destructive. Why is it that the electrons don't just zoom into the nucleus, collapsing the atom and bringing about the end of all matter?*

The problem of why atoms don't collapse concerned physicists as a gradual understanding of the structure of atoms was built up. Initially it was thought that the positive part of the atom was a bit like a positively charged jelly with the negative electrons suspended within it. (The actual image used at the time was of a Christmas pudding, known then as a plum pudding, with the electrons playing the part of raisins.) But experimental evidence showed that the positive charge was concentrated in a tiny nucleus† at the heart of the atom.

With this realisation, the common-sense parallel was the solar system (somehow, this seemed more scientific than a plum pudding). After all, gravity ensures that all the planets are attracted to the Sun and do, indeed, fall towards it. But no catastrophe ensues because the planets are also moving sideways, at 90 degrees to that inward motion. They do so at just the right speed to miss the Sun, keeping a relatively constant

* This would be distinctly embarrassing for all concerned.

† The name 'nucleus' was borrowed from biology by New Zealand-born physicist Ernest Rutherford, working at the time in Manchester. The term was already used for the central part of a eukaryotic cell, the kind of biological cell found in you and all other organisms with complex cells.

distance away – this is what is involved in being in orbit. How convenient, then, if electrons were in orbit around the nucleus too. It made for a beautiful, elegant image of consistency between the vast scales of space and the sub-microscopic scale of the atomic structure. So much so that the standard graphic representation of an atom still usually involves electrons merrily orbiting. (Take a look, for instance, at the logo of the International Atomic Energy Agency.) Which makes it a shame that the idea is a total non-starter.

IAEA logo.

Unfortunately, electrons have a habit of giving off energy in the form of electromagnetic waves when they are accelerated. That's how radios and mobile phones work. The transmitter accelerates electrons back and forth in the aerial, giving off electromagnetic radio waves which travel through the air (or through space) to the receiver. If the electrons around an atom were in orbit, they would be constantly accelerating. This may seem counterintuitive as they are not getting faster and faster, but acceleration can be a change of speed *or* of direction of travel – and to stay in orbit requires a constant change of direction. So orbiting electrons would give off energy and plummet inwards. Once again, atoms would self-destruct, and you wouldn't exist.

The solution to this was radical and slightly bonkers. Led by Danish physicist Niels Bohr, the physics community decided that, in essence, electrons would have to be restricted to particular ranges from the nucleus, not able to move inward or outward smoothly, but only in jumps, known as quantum leaps. This was one of the earliest aspects of quantum physics, which came to be the standard way to understand the (sometimes very strange) behaviour of anything very small. By restricting the electrons to these regions, as if they ran on tracks around the nucleus, disaster was averted. No one could say why this happened, it was just the way things were.

As quantum theory was developed, it became clear that the structure of atoms was more convoluted than just having electrons running along fixed tracks. Left to their own devices, particles such as electrons could not be pinned down to clear positions and trajectories. Instead, a better picture of the electron's orbital (the word devised to get away from the fixed image of an orbit) was a fuzzy cloud of probability around the atom. The electron would be somewhere in that three-dimensional space, but its exact location at any one time could not be predicted.

With electromagnetism keeping the electrons in place, another force was needed in the nucleus. Something else has to keep those quarks bound in place (and for that matter to ensure that the positively charged protons made from them don't fly away from each other due to electromagnetic repulsion). We don't see quarks at all in the wild for the simple reason that a very strong force holds them together. Unlike the familiar forces of gravity and electromagnetism, this force only operates over very small distances, but when it does apply it has the weird behaviour that as you separate two objects, the attraction between them gets stronger. In this respect, it acts more like a taut elastic band than a force like electromagnetism.

This strong nuclear force (imaginatively called 'the strong nuclear force') ensures that we never see stand-alone quarks. What's more, enough of it leaks out of a proton or neutron to hold those compound particles together in the nucleus. Because of the very short range of the strong force, there is a limit to the size of the nucleus – this is why there are only a few tens of different chemical elements, because as you continue to add more protons and neutrons, a nucleus becomes increasingly unstable.*

And that's all we need. There are other particles and other forces needed to complete the big picture of how the universe works as a whole, but to get the basics of atoms (and hence the fundamental basics of you), that extremely small number of components does the job.

THE SPACE IN BETWEEN

In one sense, it's worth noting, though, that there is a seriously important component of what you are missing from that analysis. That is nothing. This is where language lets us down. The missing component, which is by far the biggest constituent of you, is nothing. Emptiness. The void. This isn't a nihilistic plunge into the darkness of the soul, but a realistic assessment of your composition.

Let's take a zoom in on those basic components existing within the simplest type of atom in your body, a hydrogen atom. If we could somehow visualise what goes on in the quantum complexity of the sub-microscopic scale, we would expect to see somewhere in the middle a single proton, made up of two up quarks, a down quark and the gluons holding it together. Around the outside in a fuzzy ball of probability would be a single electron. And in between would be empty space – absolutely loads of nothing. The hydrogen atom is around 99.9999999999996 per cent empty space. An old simile for the relative size of the nucleus and the hydrogen atom is a fly in London's Albert Hall. Another way to look at it is that if the atom were the size of the Earth, the nucleus would be about 200 metres across – the rest of it is empty space.

* It's also why the very heavy elements, such as the most recent additions from 113 to 118, including nihonium, tennessine and oganesson, have extremely short lifetimes and limited use other than as answers on TV quiz shows.

CHEMICAL COMPONENTS OF YOU

What we've seen so far is a physicist's view. For many, a more familiar way to look at those building blocks of you is as chemical elements. This pushes the number of basic components up quite a lot from four basic particles, but bearing in mind there are about 7×10^{27} atoms in a typical 70 kg (150 lb) human, it's still quite a simplification to realise there are only 92 types of atom available naturally to play with on Earth, and of those, we can account for 99.95 per cent of your body weight with just eleven.

You've probably heard that the majority of your body is water. It might seem unlikely when you look at yourself in the mirror or rap your knuckles together. The body can feel quite solid. But bear in mind that most of you is made up of cells that are filled with water. There's enough structure to make it unlikely that you will run down the drain, but there is certainly plenty of water there. The most common figure is that around 60 per cent of your body is water – even your oh-so-solid bones are about 30 per cent aqueous.

Knowing as we do that water is made up of hydrogen and oxygen – H_2O – it might seem that this implies that the elements topping the charts for body weight should be hydrogen and oxygen, but there's also a huge amount of that most versatile of atoms, carbon, present. All life as we know it incorporates water, and is based on carbon structures. It's just about possible to consider life using other fluids to provide the environment that water does. Some have suggested, for example, there could be life which uses methane (CH_4) instead of water on the chilly surface of Saturn's moon Titan (averaging around –180°C/–290°F), where methane is liquid. But it is very difficult to see how any other atom could take over the role of carbon.

The reason is simple – carbon is by far the most versatile element when it comes to making a wide range of structures, from hexagonal benzene rings up to the huge carbon chains that make up polymers, such as those in plastics, and massive molecules such as DNA. It has been suggested in the past that silicon, which occupies an equivalent position to carbon in the periodic table, could be a replacement, giving us the possibility of silicon-based life. The idea was, at one point, beloved of science fiction writers.* But the reality is that silicon just isn't up to the job.

While it's true that chemists managed to make a short-lived, mangled equivalent of a benzene ring from silicon in 2009, to quote one of the team from Imperial College London responsible: 'What is stable and normal for carbon is unstable for silicon, and by the same token what is unstable for carbon is stable for silicon. It's an upside-down world.' The essential molecules for life with silicon substituted are simply not able to form.

Because each carbon atom has 12 times the weight of a hydrogen atom, in practice it is carbon that comes second to oxygen by weight, with about 65 per cent oxygen, 18 per cent carbon and 10.2 per cent hydrogen in your body. With those three elements alone, we've made up over 93 per cent of your body weight. Throw in a small amount of nitrogen (3.1 per cent), a pinch of calcium for those bones (1.6 per cent), 1.2 per cent phosphorous, around 0.25 per cent each of potassium and sulfur, with smaller percentages of sodium, chlorine and magnesium, and you've hit that 99.95 per cent mark. You are likely to find traces of another 40 to 50 elements, but most contribute little to your existence.

HOW MUCH IS YOUR BODY WORTH?

One way to assess just what goes into making you up at this chemical level is to look at the value of the constituents of your body on the open market. It should be stressed that this has nothing to do with your value

* You'll find a silicon-based lifeform in the episode 'The Devil in the Dark' from the original series of *Star Trek*, giving Dr McCoy the chance to complain, 'I'm a doctor, not a bricklayer.'

as an individual – or, for that matter, with the potential value of your organs and other body contents on a medical basis (some estimates put this as high as £35 million or $45 million). Here, we are considering the value of the individual elements in your body, breaking you down to your component atoms. This isn't an easy calculation, as it can be difficult to get reasonable assessments of the market price for an element, but it has been estimated to be around £125 ($160).

Such estimates vary hugely. To see how easy it is for this to be the case, consider those highly important components of your body, oxygen and hydrogen. The estimate above used a cost per kilogram of £0.17 ($0.20) for both oxygen and hydrogen. But water (just hydrogen and oxygen) costs less than this – my latest water bill gives a charge of £0.13 per kilogram, and that's not the cheapest way it could be obtained. In total, by this estimate, the hydrogen and oxygen in your body together are worth around £9.40 ($11.40), but this is far outpriced by the 160g of potassium in your makeup, which is given a value of £86 ($104), dominating your body's chemical worth.

Again, if we try to buy an equivalent amount of potassium, we get a whole range of pricings. Going for lab-quality potassium metal, for example, I'd have to pay around £414 ($500) for 160g. On the other hand, a banana contains around 0.4g of potassium – so 400 bananas would give us the same amount of potassium. I can get my 160g of potassium by buying those from a supermarket at a cost of £56 ($68). If I bought the bananas wholesale, I could halve that price.

It's clear that we're never going to get an exact value here. Other estimates put the value of the chemical makeup of the body as anything from £0.83 ($1) to £1,650 ($2,000). This seems a bizarrely extreme range. In the figure producing the high values, for example, hydrogen dominates, because it has been priced at £83 ($100) per kilo, apparently based on the price of hydrogen fuel for cars. The low value uses relatively old data and almost certainly involves a calculation error. Even so, we get a feel for the cost of what is inside you at this simple level.

THE LIFE STORY OF AN ATOM

Every one of those circa 7×10^{27} atoms has to have come from somewhere. They weren't constructed from nothing when you were conceived, or born, or as you have grown. The atoms in your body are constantly being replaced at different rates – some remain only hours, others will have been in you for a few years, but over a ten-year period the majority of them will have been replaced. And there are only two obvious sources for atoms to join your body – the air that you breathe and the food and drink that you consume. We'll cover the details of the consumption side in the next chapter, but the key consideration here is that the atoms that become incorporated in your body will previously have been in the air, plants, animals* and minerals.

If we could follow an individual atom back through its history, it will have been incorporated many times into other animals and plants. There are so many atoms involved here that just as we can say that you are descended from royalty, we can also say with certainty that your body incorporates atoms that have previously been in the bodies of royalty or the historical celebrity of your choice. Bear in mind that your body alone contains around 100,000 times more atoms than the order-of-magnitude† estimate of 100 billion (10^{11}) humans ever existing.

In fact, your atoms have been in pretty well every type of living thing, from trees to grass, insects to dogs. Go further back and we can say with certainty that the very same atoms will have been in dinosaurs, while throughout the presence of life on Earth many will also have been

....

* Sorry, vegetarians and vegans, but even if you only eat plants, many of the atoms in those plants will have previously been in animals. So will atoms in the air. You consume plenty of animals indirectly.

† Order of magnitude, a term often used in science, usually means to the nearest power of 10 – so order of magnitude 100 billion means the value is closer to 100 billion than it is to 10 billion or 1,000 billion.

in bacteria. With the exception of a few atoms produced by radioactive decay, every atom in your body will have already existed when the Earth formed around 4.5 billion years ago.

So, if you are made up of atoms, and those atoms were already in existence way back when the Earth came into existence, where are you actually from? What is your atomic heritage? The universe is around 13.8 billion years old. What happens if we trace the atoms back before the Earth and our solar system came into being? The solar system formed from space-borne gas and dust which itself could only have had two sources. The earliest of these is, effectively, the big bang.

THE COSMOLOGIST'S TIME MACHINE

We can't with 100 per cent certainty describe how the universe began. It would be unfair to be too hard on cosmologists in this respect. Compare the job of an archaeologist and a cosmologist. The archaeologist will typically be trying to deduce what happened a few thousand years ago, using artefacts that they can get their hands on and test using a whole range of tools and techniques. The cosmologist is looking back *millions* of times further into the past and can't touch or directly analyse anything.

To be fair, astronomers and cosmologists do have one advantage. When I was young there was a TV show called *The Time Tunnel* which, as the name suggests, involved a psychedelic-looking tunnel-shaped device through which the characters could peer and see past events or even pass through to visit another time.* If we limit the capabilities of a time tunnel to viewing, it is rather similar to the bonus that cosmologists have when looking into space over the archaeologist's world. Although archaeologists can touch and examine relics, they can never see directly

* In my mind, *The Time Tunnel* was a black and white show – apparently it was actually made in colour, but I only ever saw it on a black and white TV set.

how they were used. By contrast, astronomers and cosmologists can't directly interact with what they observe, but they can see into the past.

There was no meaningful explanation for the working of the time tunnel in the TV show, but our real, space-facing time tunnel functions due to a simple reality about light – it has a speed. Light moves quickly. Very quickly. A hummingbird's wings flap 4,200 times in a minute, near invisible to the human eye. Yet in the duration of just ten flaps of those wings, a beam of light could have travelled the distance of the Earth's circumference. Light is so fast that for a long time it wasn't even clear whether it had a speed at all, or whether it got from one place to another instantly. The 17th-century French philosopher René Descartes favoured instantaneous travel, thinking that light acted rather like a snooker cue. Push one end of the cue and it appears that the other end moves instantly* – so imagine light starting at a distance as a push somewhere at the other end of some intervening substance and it arrives at your eye the same moment.

Not long after, the first attempts to measure the speed of light showed that it was indeed very fast, but not instantaneous. It travels at around 300,000 kilometres (186,000 miles) per second. In fact, unlike most other constants of nature, we can put an exact figure on its velocity: 299,792,458 metres per second in a vacuum (it goes slower when it travels through a substance). It's possible to be this precise because a metre, originally defined as 1/10,000,000th of the distance from the North Pole to the equator through Paris, is now defined as 1/299,792,458th of the distance light travels in a second.

* Unfortunately, this model is itself incorrect. When you push one end of an object the other end doesn't move instantly, but rather the push, travelling from atom to atom through inter-atomic forces, moves through the object at a finite speed like a wave.

A JOURNEY INTO SPACE

There is, then, an inevitable time-shifting effect when looking up at the stars. The further an object is away, the further back in time you see it, as you have to wait for the light to get from the object to your eyes. The next clear night, see if you can spot the constellation Orion, recognisable by the distinctive belt made up of three stars.

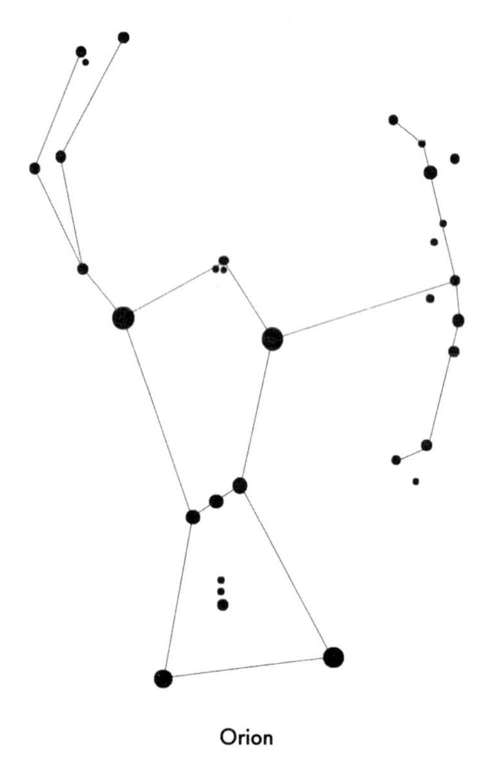

Orion

Alnilam, the middle of the three stars in the belt is around 2,000 light years away. Despite regular attempts in sci-fi movies to use 'light year' as a measure of time, it's actually a distance. As the name suggests, a light year is the distance light travels in one year, about 9.5 trillion kilometres.

So when you look up at Alnilam, the light you see has been on its way towards you for the last 2,000 years. You are seeing Alnilam as it was then, not as it is now. Other, brighter objects give us a more penetrating look into the past.

The most distant object visible to the naked eye is probably the Andromeda galaxy, visible to those with good eyesight on a clear, dark night as a little fuzzy patch relatively near the W of Cassiopeia. Andromeda is the nearest big neighbour to our own Milky Way galaxy. But even so, it is 2.5 million light years away, so we see it as it was 2.5 million years in the past, long before humans existed.

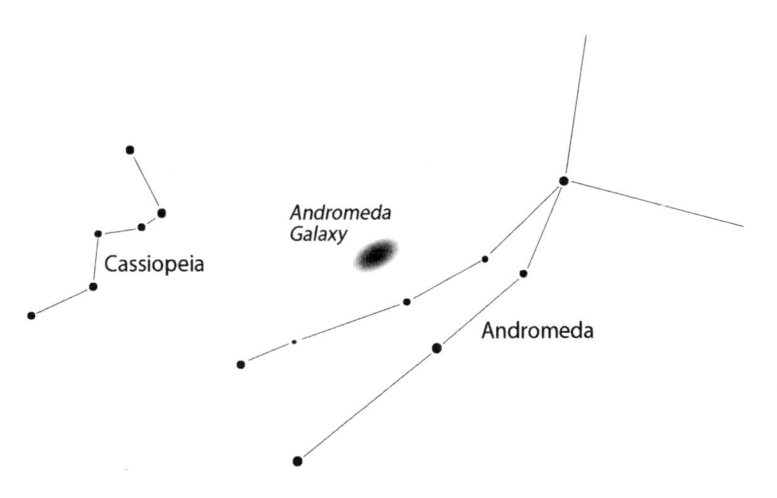

Location of the Andromeda galaxy in the night sky.

With advanced modern telescopes and by using a wide range of the light spectrum from radio through X-rays and gamma rays, rather than just the limited light capacity of our eyes, astronomers can see billions of years into the past. But such views are limited, and a lot of what we think we know about the universe 13.8 billion years in the past involves a degree of speculation, dependent as it is on a series of very indirect measures.

HOW IT ALL BEGAN

The lack of direct evidence means that the big bang theory of the origin of the universe does have competitors as a description of how things began. However, for the moment, it's the best interpretation of the data we have, and as such is the widely accepted theory.* According to the big bang theory (the actual theory, not the TV show), the universe began as an impossibly small dot of pure energy. Like the electron, it had no dimensions and, as such, gives us some theoretical issues. Shortly after its origin, though, it began to expand, and physics as we know it can describe what happened well.

Initially the stuff of the universe was so hot and energetic that there were no atoms. But physics tells us (and experiments, including the explosions of nuclear weapons, demonstrate) that energy and matter are interchangeable. When the young universe was a few minutes old, some matter, in the form of atomic nuclei, was able to come into existence. Initially all that we had was hydrogen. Lots and lots of the simplest atom of them all. Briefly, at this point in time, there was enough temperature and pressure for the whole baby universe to act as if it were a star.

Stars are factories for turning one type of atom into another. The heat and immense pressure inside a star squashes atomic nuclei together sufficiently that, with a bit of help from quantum physics, nuclei can merge together in the process known as nuclear fusion.

By far the simplest fusion process (the one that powers the hydrogen bomb, as well as most stars) is when hydrogen fuses into the next element up in mass, helium. And this happened to an extent, briefly, in that early universe. There was even a touch of fusion to the next element, lithium. But this whole process only took place for about seventeen

* This, it should be stressed, is the nature of science. Science is not capable of finding absolute truths, except in very simple circumstances. Usually it's about having the best-supported theory given the current evidence, but that theory is always subject to revision.

minutes, then the expansion and cooling had gone too far. The universe was no longer an uber-star. This meant that in the early universe, before any true stars had formed, there was mostly hydrogen, some helium and a touch of lithium.

If that was all there were to the history of the universe, you wouldn't exist. The hydrogen may be essential to make your body, but unless you've recently been making silly voices with a helium balloon, there is pretty well no helium inside you and only a tiny trace of lithium. In that early universe there would be none of those equally essential carbon and oxygen atoms, let alone all the lesser constituents, because there literally were only three elements in the whole universe. Frankly, as a place to be, it would have been extremely boring. But gravity was doing its thing and very soon in the timescales of the universe it was pulling together collections of atoms to form the earliest stars.

THE STARDUST FACTORIES

Initially, the stars that formed from the gas that filled the universe did little more than the universe itself had done in assembling helium atoms from hydrogen. But eventually other, heavier atoms began to form. Stars are not limited to getting energy from hydrogen fusion but can also produce the other atoms up to iron – number 26 in the periodic table. That, though, is the end of the line for this process known as stellar nucleosynthesis, because it would take more energy to take iron through another step to reach zinc than is available from this nuclear process.

By the time we've got iron, we're a lot better off in terms of having access to the elements that make up your body – in fact, the first trace element you would be missing is the aforementioned zinc, which amounts to just 0.003 per cent of your body. However, in terms of supporting life, we would still have the problem that those elements would be sitting inside a star, which is not exactly a comfortable environment.

Luckily for us, some of the old stars became unstable. They collapsed inwards, then exploded outwards in the vast cosmic convulsion that is a supernova.

This had two benefits. Firstly, it got the heavier atoms out there in space alongside the remaining hydrogen, so there was a cloud of gas and dust that had the potential to form planets (more on that in a moment). Secondly, that violent collapse would have pushed the atoms together far more than even the star could – enough to generate elements up to around rubidium – number 37 in the periodic table.

GETTING HEAVY

We're doing well indeed, but there are still other tiny percentages of bits and pieces (not necessarily beneficial) in your body. For those we need two further processes. One involves second-generation stars, which already include iron from a previous supernova. These stars typically vary between about half the mass of the Sun up to around ten times as heavy. Despite being relatively low mass, these stars are typically much bigger than the Sun, known as red giants. Eventually such stars can blow off their outer parts, leaving behind a small white dwarf star, sending heavier still atoms out into the universe. This process can produce elements all the way up to lead, element 82.

For the rest of the elements (which have no value in your body at all), as well as extra doses of the elements from silver (element 47) upwards, it has recently been discovered that a likely cause is neutron star mergers. Neutron stars are the endpoint of the lifecycle of some large stars (above around eight times the mass of the Sun), which have a mostly iron core and which go supernova – what is left behind is a mass of pure neutrons, which can huddle up together without all the usual intra-atom spacing, making a neutron star immensely dense. Just a teaspoonful of neutron star material would have a mass of about 5 billion tonnes.

It is common for stars to form in pairs, orbiting each other. When a pair of neutron stars are formed, they can end up close enough to eventually spiral into each other and merge, in the process blasting out these extra-heavy elements.

Over time, the clouds of dust and gas spewed out by first- and second-generation stars started to pull together under the force of gravity. Gravity, after all, goes on for ever – it has no limits. Although the pull between individual specks of dust was tiny, with nothing to resist it, in time particles started to move. The result would eventually be the formation of a rotating disc of material. It was a disc because of the rotation – rather like the ball of dough that becomes a pizza spins out into a flattish circle. The largest quantities of material congregated in the centre, eventually coming together to form a star – in our case, the Sun. Further out, accumulating dust and gas resulted in the formation of planets. In our solar system, there was enough heavy material to generate rocky inner planets as well as the primarily gaseous outer planets.

Finally, we had the opportunity for all those elements that would eventually end up in your body to do something. The very active young Earth mixed things up nicely. The heavier elements largely headed towards the centre of the planet under the pull of gravity – this is why, despite iron being the most common element in the Earth, there is relatively little on the surface. The dice were being loaded for what was to come next.

Atoms are wonderful, and without doubt they are what makes you up – but they are far from enough to make you what you are. A rock is just a collection of atoms too. When the 18th-century Swedish naturalist, Carl von Linné, better known by the Latinised version of his name, Linnaeus, first put together his classification system for the natural world making use of Latin two-part 'binomial' names such as the familiar *Homo sapiens*, he clearly had this in mind. He divided nature into three kingdoms: animals, plants and minerals. Now, though, we make a clearer distinction. Minerals have been ousted from the classification because they aren't alive.

The next thing we need to add to the mix of elements that make you *you* is life itself. Whatever that is …

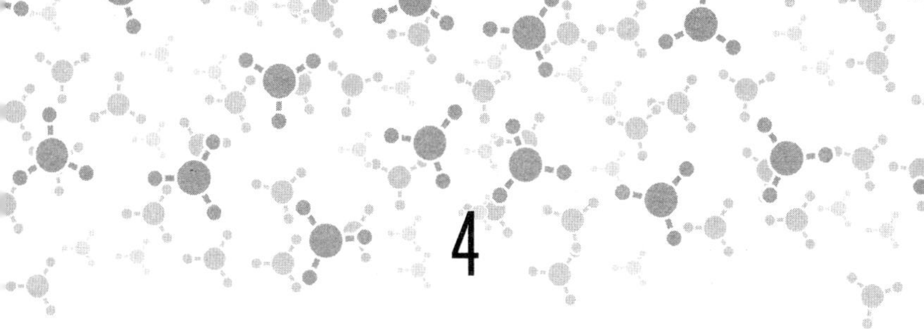

4

WHERE DID THE SPARK COME FROM?

If you were to look for the essence of what distinguishes 'you' from a collection of atoms or physical structures, it is surely that you are alive. The assorted elements that were priced up in the previous chapter comprise every physical object within you – but it's clear that the way they are arranged and interact produces something that is very different from a bag of chemical elements. Yet life is a surprisingly difficult concept to pin down.

WHAT IS LIFE, ANYWAY?

In many cases, life is one of those things that is much easier to identify than it is to describe. We can look at many objects and say straight away whether they are alive or not. It's easy enough to distinguish, say, the life present in a person, a dog and a daffodil from the lack of life in a rock, a piece of wood and a piece of plastic. Yet, of course, these distinctions aren't always obvious. Is a cut flower in a vase alive? How about a seed in a seed packet? If we take a definitely living organism – a slug, say, so we don't get too attached to it – if you cut part of it off, is that part alive? How about a single cell, the definitive sub-unit of living organisms?*

It is now possible to produce a cell culture, a collection of cells grown in the laboratory from an original cell. The best-known human example of these is the HeLa line, made famous by the book *The Immortal Life of Henrietta Lacks*. Lacks was a cervical cancer patient who died in 1951. Cells were taken from her tumour and grown in the lab,† becoming a vital resource in research on cancer and HIV. Because they are cancer cells, the HeLa line are so-called immortal cells: unlike ordinary cells in the body, they can continue to split over and over again forever, producing new cells. Over 20 tonnes of HeLa cells have now been produced.

Clearly the HeLa cells are not a person. But are they alive? The difficulty here comes down to having a rigorous enough definition of just what being alive is. Life was once considered a separate force or fluid, often associated with air. What made us alive was thought quite literally to be the 'breath of life' – reflected in the origins of the word 'spirit' and its links with, for example, the scientific term 'respiration'.

Although the link of the spirit of life with actual air died out several hundred years ago, an idea persisted of there being a life force as a hand-waving kind of energy that makes living things different from non-living. The 19th-century Scottish scientist James Clerk Maxwell as a boy was always asking 'What's the go o' that?' meaning 'What makes it go?' His questions reflect that we know in a loose way that energy is required to make things happen – it seemed natural that life had some kind of essential energy behind it, which was not present in a 'never alive' thing, and which would have departed from a 'once alive' thing.

* All living organisms are made up of cells, which are tiny bags called membranes surrounding a watery gel called cytosol which can contain a whole range of molecules and molecular structures. More on these later.

† Controversially, as was common at the time, Lacks and her relatives were not informed that the cells were being used.

The life force as a distinct (if vague) thing lives on in some traditional medicine (think of *qi* in Chinese traditional medicine, for example) and in loose descriptions such as being 'full of life' – but it is not a scientific concept. Clearly, living organisms make use of standard, common-or-garden physical energy, produced from food being chemically reacted with oxygen from air – but this is no different from the way energy is produced by a burning fire, which is certainly not alive. It is difficult to put life down to some magic additional energy as there is no actual evidence for this phenomenon existing.

Recently, science has begun to look at defining life in energetic terms (though not related to any 'life force'). We'll come back to that later in the chapter, but for the most part, biologists have been limited to identifying life by what it does. Unlike, say, a rock, living organisms do things of their own accord. Of course, 'doing things' is, of itself, nowhere near enough. The weather does things. For that matter, so do earthquakes. But neither of these is alive. Instead, life has been allocated a number of key processes – usually seven of them.

IS THIS THE REAL LIFE? IS THIS JUST PROCESSES?

The traditional seven processes of life are movement, nutrition (consuming a fuel to generate energy), respiration (the process used to harness energy), excretion (disposing of waste), reproduction, sensing (interacting with the environment) and growing. The trouble is, most of these processes can be found in non-living things, and some things that probably *are* alive don't have all these processes available to them.

The classic 'Is it, or isn't it, alive?' example is a virus. A virus doesn't satisfy all those seven criteria because it cheats by taking over the mechanism of a living cell – left to its own devices, it can't reproduce, for example. At one time, this meant viruses were solidly labelled as 'not alive', but now there's a broad feeling that just because they're hijacking another organism's mechanisms, this shouldn't exclude them.

In the 1990s, NASA came up with an alternative way of identifying life by requiring it to be a 'self-sustaining chemical system capable of Darwinian evolution'.* Is this any better? It's certainly less prescriptive, though it still leaves the virus teetering on the edge of whatever 'self-sustaining' means. In one sense, no organism is self-sustaining, as it needs to take in energy from outside – it is sustained by the energy source. That being the case, we can, perhaps give the virus the benefit of the doubt. Whatever its status, we can hopefully agree on the assertion that *you* are alive, something that was also true of your ancestors. But where and how does that chain of life start?

How life got started is a mystery that has intrigued scientists for centuries. From the early days of the Ancient Greek philosophers, one possibility was spontaneous generation, the idea that life could emerge from nowhere. Aristotle, for example, while recognising (if misunderstanding) the role of sexual reproduction, also allowed for organisms to simply spring into existence with no ancestors. Looking at his examples of spontaneous generation – for example, life coming from putrefying matter, or the insides of animals – it seems likely that Aristotle was taken in by the way that insects often lay their near-invisible eggs in such places and are then long gone before the apparently unrelated larvae 'spontaneously' appear.

It was only with more of an experimental approach to science that spontaneous generation would be dismissed. Rather than, for example, assuming that maggots were spontaneously generated on rotting meat, tests were made with containers of the same meat, some open to the air, others sealed. The maggots only appeared in the open containers, showing that there was some external source of contamination – in this case, flies laying eggs – going unnoticed.

....

* It might seem odd that a space agency should be coming up with a definition of life – a bit like a biologist designing spacecraft – but this was in the context of searching for life on other planets.

In trying to discover the nature of life, biologists tend to approach what it is from the top down, looking at organisms and trying to understand what makes them work. Physicists, though, have provided different insights into the way that flows of energy from place to place could produce life-like structures, or even help us to understand the origin of life itself.

THE LIFE THERMODYNAMIC

Generally speaking, there has been relatively little interaction between physics and biology. (It didn't help that the great physicist Ernest Rutherford once made a cutting remark aimed at biologists' focus on identification and classification that 'All science is either physics or stamp collecting'.) However, a considerable contribution to our understanding of biology was made by Austrian quantum physicist Erwin Schrödinger. In 1944, Schrödinger published a book based on a series of lectures given in his adopted home of Dublin. *What is Life?* made energy flows central to the nature of life, along with a physical phenomenon called entropy.

Entropy is a feature of the second law of thermodynamics. At its simplest, this important physical law says that heat passes from a warmer body to a cooler body if they are in contact and no energy is put into the system.* But the same law can also be phrased in a more interesting way that 'entropy in a closed system stays the same or increases'. Entropy is a measure of the amount of disorder in a system.† This sounds a

* The 'no energy is put into the system' bit is essential. If you *can* put energy into the system, it's easy to make heat flow from a cooler place to a warmer place rather than the other way round. You probably have such a remarkable device, known as a refrigerator.

† We've used the word 'system' here a lot. It just means any collection of things which interact – your body, for example is a system. A system can either be closed, in which case energy can't get in or out of it, or it can be open, in which case it exchanges energy with the outside world. Your body is a very open system. Some scientists allow for three types of system: open, closed and isolated. With that definition it's

little fuzzy as a concept, but entropy is a precise mathematical construct, reflecting the number of ways the different components of the system can be arranged.

If you think of this book as a system, with the letters in it as its components, there is only one way to arrange the letters to make up the specific book that you are reading (ignoring swapping identical letters). There are vastly more ways to rearrange those letters to make something other than this book. Accordingly, the book has a much lower entropy than has the shuffled-up set of letters. And if you imagine that all the letters were loose on the page, it would be much easier with a quick shake to go from the ordered collection of letters in the book to a disordered mess of letters than it would to go from a scrambled up set of letters to the (hopefully) meaningful book.

Living organisms require a lot of structure and order – so have lower entropy than a random collection of the atoms that make them up. You may consider your life disordered, but in reality, there's a remarkable amount of order required to make you out of those atoms. Schrödinger's way of looking at life was as something capable of doing this – of using energy to push against the standard behaviour of the second law of thermodynamics.

In *What is Life?*, Schrödinger also made the suggestion that biological inheritance would depend on an 'aperiodic crystal' – a molecule that could carry information in its structure. This would be a crystal that instead of having the simple repeating structure we're familiar with to form something like a salt crystal, has a far more complex structure that allows it to carry a whole string of information. This was a prediction that was fulfilled with the discovery of the structure of DNA. But the aspect of thermodynamics would later be revisited by physicists to suggest how life could have got started in the first place.

....

an isolated system where energy can't get in or out – a closed system only prevents matter getting in and out – but I prefer the simpler terminology.

SOUP AND LIGHTNING

Before we get onto the thermodynamic approach, we can trace a more biologically driven exploration of the origins of life from the 1950s. If you've ever seen a film of *Frankenstein*, you will be familiar with the idea of employing the vast electrical power of lightning to give the initial kick required to bring life into being. Interestingly, in Mary Shelley's original book, it's not entirely clear that electricity was used in this way – it's the 1931 James Whale version that gives us that classic image of the mad scientist harnessing the lightning bolts on a stormy night. It's hard to believe that the movie wasn't at least part of the inspiration for an experiment devised by Stanley Miller, a PhD student working at the University of Chicago in the lab of chemist Harold Urey.

The aim of Miller's experiment was to create a miniature replica of the conditions that existed on the early Earth in a spherical glass container that would not have looked out of place on the *Frankenstein* set. Miller zapped electric sparks across the vessel which contained water vapour, methane, ammonia and hydrogen, assumed to reflect the Earth's atmosphere when life was thought to have begun. The electrical discharges stood in for the lightning that was thought to be a regular feature of the early landscape and that it was hoped had provided the energy to kick-start life.

After the experiment had run for a week, the contents were analysed, and the results seemed impressive. Miller had not created life, but a number of chemical reactions had occurred, producing a soup of organic molecules. We need to be a little careful here with this word 'organic'. Putting aside the marketing use to denote goods produced in a particular way, we tend to associate the word 'organic' with life. However, in chemistry, organic simply means a compound contains carbon – so methane, for example, one of the ingredients for the experiment, was an organic compound.

The molecules produced, though, were more complex than methane (which is just a carbon atom with four hydrogen atoms attached),

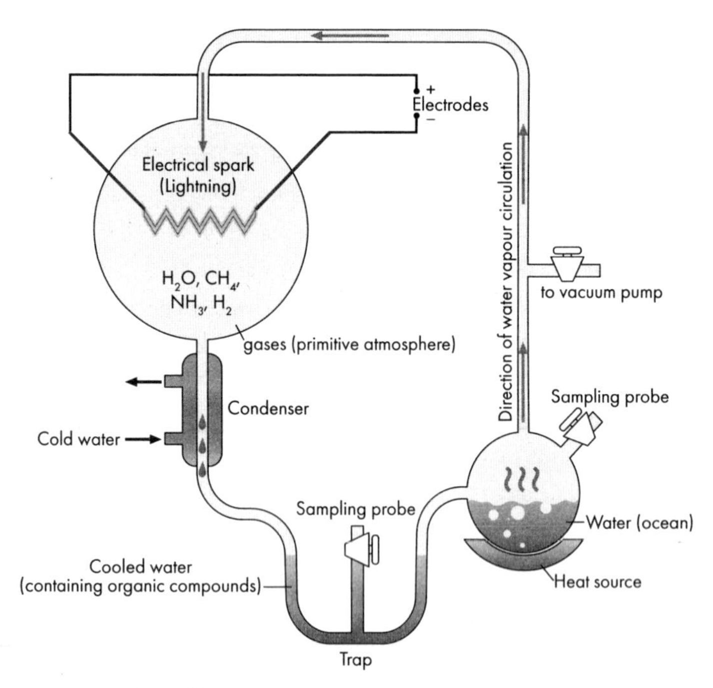

The Miller–Urey experiment attempted to recreate the environment when life started.

most notably in an amino acid called glycine. Amino acids are acidic organic compounds containing nitrogen, twenty of which can be specified by the genetic code of DNA on the way to building the proteins that are essential for life. Glycine is one of these twenty compounds. At the time, this was seen as a clear indicator that life could have been electrified into existence out of the common gases present in the early atmosphere. The outcome of the experiment seemed to gain support in 2007, when a sealed sample of the original mix was retested and discovered to contain a number of other amino acids that had not previously been detected.

The trouble with the Miller experiment is that the conclusions drawn from it are not truly scientific – it might feel right, but it's hard to follow a logical path from the outcome of the experiment to life being started. As a result, the findings are now considered fatally flawed. There was one aspect that has always been a problem. Amino acids are certainly necessary for life as we know it – but they are very low-level components compared with the complexity of the simplest living organism. Amino acids form quite easily – they exist in some regions of space, for example. Making the leap from having generated a few amino acids to producing life is a little like saying that once we have produced a box of gears and bolts, we have built a car. In each case, we have some basic components, but nothing that resembles the final product.

Things proved even less favourable for the whole Miller picture, though, because an essential assumption that was made in designing the experiment was incorrect. That mix of water vapour, methane, ammonia and hydrogen has been supplanted by a different makeup for the likely early atmosphere as we get to know the nature of the young Earth better. It appears that the atmosphere was far closer to the present one (though with gases in different proportions and lacking free oxygen): nitrogen, carbon dioxide and water vapour, with some added sulfur dioxide.

Such an atmosphere would be less likely to react under the influence of lightning – and even if it did and somehow life got started, that isn't all that's required. Living things need a continuing source of energy, which isn't going to be provided by sporadic lightning flashes. It's also worth noting particularly that lack of free oxygen. The element was present in compounds, but not on its own in what we would regard as a useful form. It would take the work of organisms that excrete oxygen as a waste product to get the Earth into a state where you could possibly exist.

CONSULTING THE CRYSTAL BALL

We can't, of course, be 100 per cent certain just when life did begin. Our best indications at the moment come from crystals of zirconium silicate, known as zircons.* It's thanks to zircons that we now know about the composition of the Earth's atmosphere 4 billion years ago. The atmospheric gases would have been largely produced by volcanoes, and some of the magma from volcanoes that erupted back then is still found in zircons, crystals which are particularly resistant to erosion. Some zircon crystals date back to around 400 million years after Earth formed – we are able to date them thanks to the decay rate of the radioactive contaminants that are partially responsible for their colouration.

Zircons are good at trapping particles from around them as they form, including cerium oxide, which can come in two forms: the balance between the two types gives a guide to the gases that had been released by magma into the atmosphere. Scientists from Rensselaer Polytechnic Institute in New York have shown that compounds of oxygen with carbon, hydrogen and sulfur dominated the atmospheric composition. That was only the start for zircon's ability to open up the past.

Zircon crystals have also provided scientists with an idea of when life began. We can only observe early life so far with fossils. But we can indirectly make use of other zircon-trapped material. Carbon has two main isotopes, carbon 12 and carbon 13 (there's also the carbon 14 used in radiocarbon dating, but that is relatively uncommon). Biological processes slightly favour the smaller carbon 12 atoms, so life tends to have rather more carbon 12 to carbon 13 than the natural ratio of around 99 to 1. The carbon trapped in zircons from 4.1 billion years ago, discovered in Western Australia's Jack Hills, has more carbon 12 than would be expected. It's likely, though not certain, that this is because there was already life present when these zircons were formed.

* Zircons should not be confused with cubic zirconia, the popular diamond substitute, which are zirconium dioxide crystals. Zircons are also used as gemstones, and particularly pure examples have been used as diamond substitutes, but impurities usually give zircons a range of colours from yellow and brown to blue and green.

THE KIT OF LIFE

So how did it all start? It's a bit like making a watch from a kit. In that scenario you need the parts, the appropriate tools, the instructions on how it all goes together, the energy to assemble it and to make it operate, and a case to put it in. For life, the equivalent of the parts is the carbon and other chemical constituents necessary to construct an organism. Just as the watch builder needs energy both for construction and to keep the clock going, so does the organism. A mechanical watch needs oil to enable smooth movement – perhaps the closest parallel in life are catalysts, substances which vastly accelerate chemical reactions, essential for most life processes. The equivalent of the clock case is something to give structure to what otherwise would be just a collection of free-floating molecules – in the case of cells, a membrane.

In two final cases, the watch is more different from our DIY life-form. As we've seen, life produces waste. In fact, as we will see in a couple of pages' time, this is not just an unfortunate inefficiency, but an essential requirement arising from the second law of thermodynamics. This applies to the watch too, it's just that the waste is not so obvious as it's generated as heat and sound rather than biological excreta. But one thing watches don't do is reproduce. Although in principle you could have life that didn't reproduce, life as we know it requires this, which means that the instructions for construction, rather than being external and separate as was the case in our watch-building analogy, need to be built into the organism itself and capable of being passed on to future generations. This is the role fulfilled in life on Earth by DNA.

What's difficult to imagine is all these things randomly coming together at the same time to suddenly enable life to emerge. The choice of making a watch in my example above was not an arbitrary one. I did so because one of the most famous arguments for creation by a deity made use of a watch. This was the watchmaker analogy used by William

Paley, an 18th-century English clergyman. Paley's argument, in the opening of his book *Natural Theology*, went like this:

> In crossing a heath, suppose I pitched my foot against a stone, and were asked how the stone came to be there; I might possibly answer, that, for anything I knew to the contrary, it had lain there forever: nor would it perhaps be very easy to show the absurdity of this answer. But suppose I had found a watch upon the ground, and it should be inquired how the watch happened to be in that place; I should hardly think of the answer I had before given, that for anything I knew, the watch might have always been there. Yet why should not this answer serve for the watch, as well as for the stone? Why is it not as admissible in the second case as in the first? For this reason, and for no other, viz. that, when we come to inspect the watch, we perceive (what we could not discover in the stone) that its several parts are framed and put together for a purpose, e.g. that they are so formed and adjusted as to produce motion, and that motion so regulated as to point out the hour of the day; that, if the several parts had been differently shaped from what they are, of a different size from what they are, or placed after any other manner, or in any other order, than that in which they are placed, either no motion at all would have been carried on in the machine, or none which would have answered the use, that is now served by it.*

This isn't a stupid argument. Paley's logic tends to be rapidly dismissed now as a result of the idea so eloquently put forward by Richard Dawkins in his book *The Blind Watchmaker* that all we need is evolution and a lot of time, and we can end up with that biological equivalent of a watch, the living organism. However, it's hard to see how all those requirements

* They don't write sentences like this one anymore.

to get our first life going could all come together simultaneously – and there's no point fulfilling just some of the requirements, as is possible with evolutionary development, the way living things could then evolve. Either we've got life, or we haven't. You can't be a bit alive.

However, we don't need to evoke a deity to solve the problem as Paley did. What we can do instead is to cheat. If we can't see a way to evolve all the requirements for life to start simultaneously, what if, instead, we can substitute naturally occurring alternatives for some of the requirements, so there is less complexity required to get things going? One strong possibility came out of the discovery of life existing in abundance where it was never expected to, deep under the oceans where there was practically no sunlight to provide energy.

THE VENTS OF CREATION

In deep water, living things have been found thriving on the energy that comes from hydrothermal vents – some known as black smokers. These vents pour out chemical-rich, extremely hot water, over 2,000 metres beneath the ocean surface. Because of the pressure at this depth, the water can be as hot as 400°C (750°F) but does not boil. Such intense heat might seem incompatible with life, but the vents are a happy hunting ground for species of bacteria which get energy from the heat and consume minerals produced by the vents.

As well as energy and raw materials, vents could also in principle provide the essential structure to contain a living organism, before membranes evolved to enable life to move away from its starting point. Black smokers can form huge chimney-like structures tens of metres tall, but alkaline hydrothermal vents produce extremely narrow structures that could act as an external support, providing a narrow space to contain the components of the living entity before it evolved a structure of its own. As a double benefit, the constant flow of water through the restricted passage could also provide a means of removing waste before any built-in mechanism took over. And the minerals produced by hydrothermal vents include metal sulphides, which can act as primitive catalysts to speed up the reactions needed for life.

Alkaline hydrothermal vents operate below 100°C and feature a network of tiny openings rather than huge chimneys, offering a potential model for a starting place for life. What's more, the alkaline nature of the vents makes it easier for a reaction of hydrogen and carbon dioxide to take place, producing basic organic structures. There's still quite a leap to be made from no life to a living thing, but a hydrothermal vent is very helpful in providing enough of the requirements for life to make the step required to bring life into existence lower. It's as if our watch kit, instead of providing us with pieces of brass, out of which we can machine gears and structures, comes with a pre-constructed movement and the parts of the case, and all we have to do is put them together.

Although these biological theories are useful to get a feel of, for example, how the natural scaffolding of a hydrothermal vent might have made it easier for life processes to start, it still doesn't get us to where life itself came from. We only have evidence of one start to life on Earth – all living organisms we have ever discovered are related – so this is the beginning of a 4-billion-year process leading to you. It is in trying to discover how life managed to first come into existence that we return to Schrödinger's physics-based approach, championed by the American physicist Jeremy England, based at the Massachusetts Institute of Technology.

THE DISSIPATIVE LIFE

England takes on life from the viewpoint that there are certain physical necessities for life to exist, which can be explored through our understanding of thermodynamics. His approach is to break down some of the well-known processes of life, such as replication, into their thermodynamic consequences to see how life-like processes could have originated. Specifically, England makes use of a concept called 'dissipative adaptation.'

As Schrödinger had pointed out, all living things dissipate energy, mostly in the form of heat, into their surroundings, increasing the entropy of those surroundings. This is necessary for structures to be formed and to be able to change in response to the environment (the process known as adaptation), an essential aspect of life. At the heart of England's approach is the idea that by increasing the entropy of its surroundings, an evolving structure is able to stay in a non-equilibrium state.

Non-living things, left to their own devices, come into a state of equilibrium* with their surroundings. This means in thermodynamic terms that the heat flowing into and out of the object balances out. There's no net heat transfer in either direction. So, for example, if you leave a hot cup of coffee on the table, over time the liquid will reach equilibrium with its surroundings. The coffee will cool down significantly, while the air around it (and the cup and table) will heat up a tiny bit, until any flow of heat between the liquid and its surroundings is balanced out by a flow from the surroundings into the liquid. They will end up at the same temperature – in equilibrium.

Life is different, though – and perhaps this gives us the best definition of what being alive really means. Living things aren't in equilibrium with everything around them, but instead take in energy from outside sources (food and sunlight, for example) and push energy out into the world, often in the form of heat. As a result of this, the organism can reduce its entropy where otherwise, naturally, entropy would increase. The living thing can grow, produce sophisticated structures and so on, up to the point where it dies and rising entropy kicks in, returning the organism to equilibrium.

* Literally 'equally balanced'.

What Jeremy England has studied is not life itself, but natural, non-living phenomena which exhibit some of this non-equilibrium behaviour, giving them a partial similarity to the abilities of living things. Two natural structures that form in this way are snowflakes and sand dunes, both of which give off energy to their environment in the process of forming. In the case of sand dunes, the grains of sand are moved by air currents, but as they move they collide with each other, losing energy to their surroundings as heat. For snowflakes there's a physical chemistry process: when water forms ice crystals it gives off heat energy as the bonds between water molecules form.

No one is suggesting that snowflakes and sand dunes are directly related to life, but England believes that there are some similarities in the way that these structures are formed, interacting with their environments, compared with the way that living things store information in their DNA. By running simulations of chemical reactions, England's team has also shown other life-like behaviour in a virtual soup of interacting materials, where for example some chemicals take over the environment by natural selection as they turn out to be better able to use the available energy.

Nick Lane, Professor of Evolutionary Biology at University College London, points out the limits of trying to discover how life started purely from a thermodynamics approach:

> I think it is possible to go too far with thermodynamics ... there has to be a selective mechanism possible. For example, if life was fundamentally about maximising entropy, rather than natural selection, then one might think that animals, or indeed eukaryotic cells [cells with nuclei surrounded by membranes, such as our own], both of which are very good at increasing entropy, would have emerged much earlier than they did – there were billions of years delay. At a broad brush-stroke level there just was not an easy selective route to that level of complexity, hence entropy could not be 'maximised' for

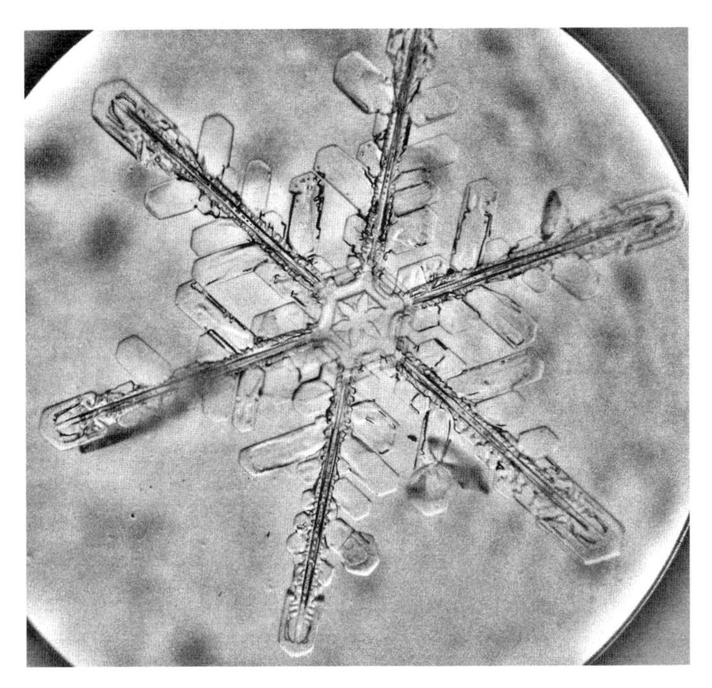

The structures of snowflakes and sand dunes are formed by energy interactions with their surroundings.

> a long time, and for that reason I would not see this thermodynamic driver as dominating the history of life.

However, it's not unreasonable to see this new viewpoint providing a more comprehensive 'starting toolbox' for life to form. And although Lane does not think we can put too much weight on the pure physics side, he points out that in his own work, modelling the way that the earliest types of cells could have formed, it is thermodynamic effects that result in structures developing in the first place. Here, thermodynamics – in his model driven by heat from hydrothermal vents on the ocean floor, which sustain a non-equilibrium system – should favour self-replication.

THE COMPLEXITY OF YOU

Of course, there's a long way to go from an initial spark of life to you. For the first 2 billion years or so that life existed on Earth, all life forms were much simpler than the perhaps 40 trillion complex cells that go to make you up. Such simple life still exists in profusion, numerically speaking dominating life on Earth, most familiarly in the form of bacteria. A bacterium is a single cell with a far simpler interior structure than is the case with your cells. But that doesn't make bacteria unsuccessful as a form of life – your body contains somewhere in the region of 50 trillion to 100 trillion bacteria, and that's a tiny proportion of the Earth's whole population.

Getting from these much simpler single-celled organisms to a complex single-celled organism would be the first step on the route to you – but that is as big a leap as the beginning of life in the first place. Our kind of cells, known as eukaryotic* cells, with a nucleus as an enclosed central 'command centre' and lots of extra molecular machinery on

* The name means, roughly, 'true kernel' in Greek, referring to the nucleus, to distinguish them from the prokaryotes or 'before kernel' organisms such as bacteria.

board, are far more complex than those in early forms of life and current bacteria. It's thought that the starting point of eukaryotic cells was a spot of symbiosis.

This is the process where two living organisms help each other for mutual benefit, sometimes losing out in other ways to help the other. As well as bacteria, there is another branch of simple prokaryotic organisms called archaea. The name implies they are the oldest living form, though there is nothing to suggest they didn't develop in parallel with bacteria. It is thought that at some point an archaeon absorbed a bacterium – but rather than the archaeon eating it, the bacterium stayed alive within the other organism. The bacterium provided energy to the archaeon in return for itself being kept alive. Over time, the bacterium's descendants ceased to be individual organisms and ended up as a functional part of a compound organism. The ex-bacteria became mitochondria (see page 17) and the enhanced archaea become eukaryotes.

After this transformation, the augmented organisms seem to have developed a complexity that has never occurred in individual bacteria or archaea, eventually making this species the ancestor of all eukaryotic organisms, including you. Like the origin of life itself, there is only evidence of this happening once. Every one of the complex-celled organisms – all animals and plants, as well as all fungi and algae – are related and originated from this one-off ancestor. Without the archaeon and bacterium getting together, you would never have existed.

IT CAME FROM OUTER SPACE

It's out of fashion now, but it is worth mentioning that there has occasionally been the suggestion that life did not originate on Earth at all, but came here from space. This isn't quite as much of a science fiction proposition as it first seems. There are a lot of places out in the universe for life to start, there are certainly basic organic molecules out there, and it is possible for some basic forms of life to survive travelling through space and crashing into the Earth.*

This thesis, known as 'panspermia', was relatively popular towards the end of the 19th century and was picked up and developed in the 20th century by two important (if maverick) scientists, Fred Hoyle and Chandra Wickramasinghe. They also suggested that a number of epidemics were triggered by arrivals from outer space. There is no good evidence for panspermia, and it doesn't solve the 'how did life start' problem, but simply pushes it to a different location. Nonetheless, panspermia is not infeasible and should not be entirely dismissed as a long-shot possibility of where your evolutionary ancestors came from.

Whichever way life first emerged, it began a long chain of existence, leading to the evolution of *Homo sapiens* and eventually to you. We always have to be careful with words in this type of topic. 'Leading to' can be taken to mean some kind of direction, as if humans were a goal that evolution has strived towards. But this 'leading to' is more in the sense of a ball's path through a pinball machine leading to wherever it ends up – in this case with you. And to get to that point needs the input of some essentials. You wouldn't be here for long were it not for the things that you eat and drink.

* In April 2019, the Israeli spacecraft Beresheet crash-landed on the Moon. The craft was carrying thousands of tardigrades, tiny organisms under a millimetre in length, sometimes called water bears. Able to withstand extremes of heat and cold in a kind of dehydrated suspended animation, all the indications are that the lunar tardigrades would survive were they recovered and rehydrated.

5

YOU ARE WHAT YOU EAT

According to the motto of Winchester College (and New College, Oxford), 'manners makyth man'. Apart from observing how delightful the medieval word 'makyth' is, it would be a more realistic proposition to suggest that water and energy play an essential role when it comes to makything human beings, you included.

THERE CAN'T BE A WATER SHORTAGE

We are all aware of the devastating impact of drought. The human body is so dependent on water that it can only go a few days without it. We are constantly losing water as vapour in our breath and through sweat and excretion, yet it is so important for the survival of our cells, and hence organs, that its absence will kill us long before a lack of food does. Water shortages commonly cause disastrous circumstances for individuals and communities. Yet, addressed logically, a water shortage on Earth is an inconceivable idea.

At the time of writing there are around 7.7 billion people alive on Earth. That's a lot – the number has more than doubled in my lifetime. But it's a number that seems suddenly much smaller when put in comparison with the amount of water there is out there. Before we look at numbers, just think of the picture of Earth as seen from space. It's not for nothing that we're described as 'the blue planet'. Water is very apparent, both in oceans and in the clouds. In total there are around 1.4 billion cubic kilometres of water on Earth.*

It's hard to relate to that, but that is around 0.185 cubic kilometres of water for each person. That's 185,000,000,000 litres of water each. We typically need around 2 litres of water a day (of which half, in practice, comes from our food) – so it would take around 250 million years to get through that water – and that would only be if we consumed the water, but didn't return it to the environment. In practice, an adult's body holds around 45 litres of water and isn't going to soak up more. The rest gets recycled.

You may hear that our water consumption, particularly in the West, is a lot higher than the volume we drink – and this is true. We each of us result in between 5,000 and 10,000 litres a day being used – in part through our activities and even more so as a result of the products we consume. It has been estimated that it takes 3,000 litres of water (fed to cattle) to make a hamburger and a remarkable 20,000 litres of water to produce 1 kg of coffee. Again, though, we have to be a little careful about this – that kilogram of coffee *contains* hardly any water. We are again talking about the water required by the system, the majority of which will be naturally recycled.

It seems, then, that there is a vast amount of water out there for every individual, and even the relatively small amount of it that we 'use' is rapidly returned. Which makes it clear that simply saying that water is a scarce commodity is just plain wrong. The problem is not a shortage of water, but that many of us live in places where water isn't readily available, and the vast bulk of the water in the world (around 97 per cent) is salt water, rather than drinkable fresh water. Both of these

* Admittedly 1.4 billion cubic kilometres is small compared to the circa 1-trillion-cubic-kilometre volume of the Earth. The thin layer of water on top amounts to around 1/700th of the total. But it's still a lot of water.

problems can be overcome – but it takes energy to do so. It's arguably energy that's in short supply.

ENERGY – WHATEVER THAT IS

This brings us neatly around to the other requirement to making you function as a biological machine. Energy is one of those terms from physics, where we all kind of know what we mean when we use the word, but it is really difficult to describe what it actually is. The great 20th-century American physicist Richard Feynman rather depressingly said that 'in physics today, we have no knowledge of what energy *is*' – and things haven't changed since. When we employ the word in everyday use, it tends to be a loose concept of the 'oomph' that makes things happen. Energy is often about motion, whether it's the kinetic energy of a moving vehicle that can do so much damage in a collision, or heat energy, which we experience through the movement of the individual atoms and molecules that make up the matter around us.

We also speak of 'potential energy'. This is really just energy that is stored up in a whole range of possible ways. The most familiar from school is probably the potential energy due to gravity. If we push a boulder to the top of a hill, then it takes work to get it up there – transferring energy to the boulder because of its position. Once at the top, we have stored up some of the energy we exerted (the rest will mostly have gone to heat), which can then be released when we let it go and it rolls back down.

That's the simplest form of potential energy – but at first glance there are lots of others. For example, the electrical energy stored in a battery that keeps your phone going, or the energy stored up in a spring when it's wound. Most importantly as far as keeping you operating, there is the energy in food. As it happens, though operating in different ways, pretty much all three of these kinds of potential energy rely on chemical energy – the energy in the bonds between atoms in chemical compounds – which in the end is a form of electromagnetic energy.*

In the battery, chemical energy in the battery's component parts is used to drive electrons around a circuit. In the spring, the electromagnetic bonds between atoms are stretched – this takes energy to do, storing it up to later release it when the bonds return to their normal state. And in our food, just as is the case when we burn something, the energy stored in the bonds between atoms is released as the bonds are broken. Typically, the digestive system breaks down your food into smaller molecules which then undergo a process known as respiration.

THE SLOW BURN

Respiration is a slow form of combustion. We usually associate combustion with flames, which thankfully aren't involved in the digestive process. But combustion is simply a matter of a fuel and an oxidising material, often the oxygen in the air, undergoing a chemical reaction whereby bonds are broken and energy released. These 'bonds' are not physical ties as we would expect in the usual use of the word, but something closer to the way that two magnets stick to each other.

As we have seen, the atoms that make up all matter – including your food – have a positive electrically charged nucleus surrounded by a cloud of negative electrically charged electrons. When two or more atoms bind together, they either share electrons between them, or one atom loses one or more electrons while another gains one or more. In

* There is one distinct and quite surprising exception to the way that these kinds of potential energy are related to chemical energy: that's in the energy stored up when you stretch an elastic band. When these strips of rubber material are stretched, the naturally kinked molecules that make them up are straightened. These molecules (like all others) naturally vibrate and jiggle around thanks to thermal energy – heat. As a result, the molecules bash into each other, pushing themselves back into the kinked form. So, it's thermal energy, not electromagnetism that enables an elastic band to store potential energy.

the latter case, the result is that one 'ion' (atoms that have gained or lost electrons) is positively charged and the other negatively charged. These particles with opposite electrical charges attract each other.

Such chemical bonds can store energy away that is released when the bonds are broken. In your body this is happening all the time. Molecules derived from your food are combined with oxygen and the energy produced is stored in a special molecule called adenosine triphosphate (ATP), which is used to transfer the energy around the body to wherever it is needed to power everything from your brain to your muscles. This is where mitochondria come in.

As we have seen, mitochondria, sometimes called the power sources of the body, are thought to once have been independent bacteria that were incorporated into many of the cells of eukaryotes. Mitochondria react fuel molecules with oxygen and produce vast quantities of the ATP molecule, which are repeatedly recycled. This is such a major process in your body that you get through around your own bodyweight in ATP every day. The chemical energy of respiration is used to push protons – hydrogen nuclei – through a membrane, setting up an electrical gradient. As a result, the molecular machines in the mitochondria can make use of there being more electrical charge on one side of the membrane than the other to power their manufacture of ATP.

MEASURING OUR ENERGY

As we're thinking of food here as our primary source of energy, it's worth exploring the units we use to measure energy in, because they are particularly confusing when applied to food. The standard scientific unit of energy is the joule. An electrical device rated at one watt uses one joule each second. The watt is a unit of power – the rate at which energy is transferred. This results in the rather bizarre unit used by energy companies of kilowatt hours. One kilowatt hour is a thousand watts for one hour – which is 3,600 seconds – so is 3,600,000 joules.

Measuring energy in kilowatt hours makes as much sense as measuring distance in 'miles per hour hours'.

The good news is that we don't measure food energy in kilowatt hours. The bad news is that it is rarely talked about in joules either. Food energy tends to be measured in calories, the scientific unit prior to the joule being introduced. Each calorie is the equivalent of 4.184 joules. As the joule is a small unit of energy, when food energy is displayed on packaging in joules, it's usually as kilojoules, represented as kJ – thousands of joules. Similarly, when displayed as calories, kilocalories are used. However, to make things really confusing, this usage came in before we all became familiar with 'kilo' style prefixes, so kilocalories are habitually mislabelled as Calories – calories with a large C.*

Getting an idea of the energy involved in food is important because, just as is the case with water, there is plenty of energy around on the Earth, but most of it is either not in the right place, or not in a form in which it can be easily used. Plants, in many respects, are better off than we are in this regard. They can get their energy directly from the Sun; with the exception of tidal, nuclear and geothermal energy, pretty well all the usable energy on our planet comes from the Sun.

We're used to the Sun as a source of light – but without solar energy we doubly would not be alive. The Sun keeps temperatures on the Earth in the right range for liquid water to exist most of the time, and it provides the energy that plants need through photosynthesis. Here, energy in the photons of light is converted into chemical energy in special reactors within the plant. Not only does this keep the plant alive, it provides the chemical energy stores which will finally reach us, either by our eating the plant directly or by our eating another organism that itself got energy from the plant.

---

* Except, food manufacturers often forget this and call kilocalories 'calories', which is simply wrong.

So, what are the energy needs to keep your body going? The average human needs around 9,800 kJ (2,350 kcal) a day, though there is considerable variation depending on your lifestyle, age and gender. A typical adult male undertaking moderate regular exercise requires around 10,450 kJ (2,500 kcal), while a typical adult female with a similar exercise routine needs around 8,400 kJ (2,000 kcal). On average we consume a higher 10,500 kJ (2,680 kcal) a day, but the total varies widely by country. The typical American, for example, consumes around 15,000 kJ (3,600 kcal) a day – an intake level that is only a good idea if undertaking extreme levels of physical exercise.

FOOD IS PLENTIFUL

The good news is that there is plenty of food out there. On average across the world, we produce enough food to provide around 25,000 kJ (6,000 kcal) a day for everyone – far more than is needed. But there are three big barriers that result in the kind of food shortages that all too often appear on the news. Like water, the food is often not where it is needed, a significant amount of it gets wasted, and getting on for half of it isn't consumed by humans. In a sense, like water waste, nothing actually gets lost. The food's atoms continue to exist and can be rebuilt as new food – all food is recyclable. But that takes time for the wasted food to be returned to the soil and atmosphere, and for new crops to grow. As a result, wastage can be an issue.

Of the 25,000 kJ total, around 14 per cent is used to make biofuels. Whether or not this is a good idea is complex. Biofuels seem to make sense from a climate change perspective (more on climate change in a moment). These are crops used to make fuel. The result is relatively green, as all the carbon in the crop will have been taken out of the air – so they are better for the environment than using coal, oil or gas. (They aren't totally green because greenhouse gases are produced by the fertiliser used and in farming, shipping and processing the crop.)

Unfortunately, all too often biofuels are edible crops, or are grown on land that could have been used to produce crops to feed people. Even if that isn't the case, the benefits of biofuels are limited, because they are an inefficient way to harness the Sun's energy. An electric car could get 200 times the range out of the solar energy from a field as a biofuel car could from the fuel produced by the same area. However, biofuels are probably worthwhile for uses such as aircraft where electric power is impractical.

A bigger, 30 per cent of the total food crop production goes to feeding animals. This isn't all wasted, of course. Those animals will primarily be eaten or, if egg- or dairy-producing, will produce foodstuffs themselves. It does mean, though, that by putting another link in the chain from the Sun to food energy to you, the amount of energy transmitted will be significantly less. On average you get around one-third of the food energy of the crop from eating meat that is fed on it compared to what you would get from the crop itself.* By far the worst translator of that food energy from plant to us is beef. Of course, not all crops fed to animals are suitable for human consumption and animals can consume crops from land where it is impractical to plant human food, such as sheep on hill farms. This means there's a balance to be struck here, and anyone suggesting that you should entirely abandon meat for environmental reasons is operating on emotional rather than logical grounds.

What about the wastage? We get a lot of coverage in news media about food that we throw away, and it's a sensible point. We are far too enthusiastic to bin food as soon as it reaches its sell-by date – an arbitrary date set by the retailer to get it off the shelf and rarely the same as a safe eat-by date, especially with fruit and vegetables which will often stay edible for many days afterwards. It's a real win-win if we don't throw away food unnecessarily, as it both avoids waste and saves us

* By eating animals, you actually get about 10 per cent on average of the food energy that animals consume, but this is improved to around one-third as animals eat things we can't, such as grass.

money. Around 4 per cent of the total available food energy is lost this way. Rather more – up to 7 per cent – is lost during distribution and processing too.

Then there is that difference between what most of us eat and what we need. While some argue we should only eat what we need and no more, it's hard not to see this as unnecessarily prescriptive. For our own good, it's best to keep close to what we need, but to suggest we shouldn't ever have an enjoyable blowout of food is a killjoy notion. Nevertheless, we would reduce what is effectively a form of waste if we (particularly those of us on Western diets) kept closer to our daily requirement most of the time – and we would be healthier too.

FEEDING THE WORLD

Of the three barriers mentioned above, the one we haven't dealt with yet is the trickiest one. The food that simply isn't where it's needed, and doesn't get there for economic, political and energy-based reasons (which feed back into the economic factor: it costs money to use the energy to ship food from a crop-rich environment to a crop-poor environment). Although aid is necessary in emergencies, it is far better if areas that have economic problems in accessing food are helped to improve their economies – but the most intractable issue tends to be political, where governments or insurgents prevent trade, restrict access to external markets or take aggressive action that turns a region into a war zone, making food movement near-impossible.

What's clear from the energy figures is that it isn't a matter of there not being enough food out there. Even with the estimated peak world population of around 11 billion, expected to be reached around 2100, after which population is predicted to tail off, this should not be a problem. However, because of the limitations we've discussed, we should be doing more to ensure that food is more cheaply and readily available. One problem here is the green movement's resistance to genetically modified (GM) crops. These can both increase yield and provide more nutrition from the same amount of food. We have been genetically modifying crops ever since agriculture began, although the mechanisms for GM are different. Of course, there need to be controls, but outright bans, such as that imposed by the European Union, are hurting those who could benefit hugely from better, more readily available food.

Getting sufficient food to the right places deals with your internal energy needs – the energy required to actually keep your body working. However, particularly in our modern world, you are also likely to have a much wider requirement for energy to provide heat and light, transport, communication (including the internet), entertainment (ditto) and more. We can't sensibly consider your energy consumption requirements without addressing this aspect of energy.

When we do so, there is a particular elephant in the room – climate change. Despite the claims of some politicians and industry lobbies, the scientific consensus is extremely solid on the facts of climate change. It exists, it is driven primarily by human activity, and unchecked it will result in widespread disasters around the world, though the exact pace of the worsening impact is certainly subject to argument. If you 'don't believe' in climate change, it is going to be difficult to persuade you. To deny the existence of man-made change with all the available evidence means logic has failed. But please read on anyway. Because energy is central to your existence, and climate change will increasingly influence what it is to be *you*.

CLIMATE CHANGE IS HAPPENING

Since the year 2000 there has only been one year when the global average June temperature has not been above what came before (2014 was the odd one out) – and 2019 continued the trend by being significantly higher than any previous year. Extreme weather events are increasing in frequency. It's very difficult to ascribe any particular event to climate change, but it becomes harder and harder to blame event after event on freak conditions – once they become the norm, they aren't oddities, they reflect the climate.

You will always get people confusing the weather and the climate. Weather is what we experience on a day-to-day basis, and the fact that, for example, we get snow, or a cold winter, says nothing about the climate as a whole. In reality, climate change is significantly more complex than just 'everything is getting warmer' (which is why the expression 'global warming' is used less now, as it causes confusion). Climate change will certainly see average temperatures increasing, as those June figures make clear, but it also implies more heavy rainfall, more droughts, and more heavy winters.

In terms of the impact on your everyday life, in northern Europe and North America we are relatively lucky. Climate change could be marginally beneficial in some regions, giving northern Europe, for example, a climate that is more like southern Europe – but even here there are some issues. For example, deaths from heatwaves are increasing, while diseases such as malaria, which was wiped out in the UK by a series of cold summers in the 1800s, could return. For other areas of the world, even a couple of degrees rise in average temperature can result in sustained drought, uncontrolled wildfires, devastating cyclones and far more.

The impact that will be most widespread is likely to be sea level rise. There is a double contribution to this. The biggest contribution right now is that seawater is expanding as those average temperatures rise, but the poster child of climate change is ice melting. At the moment this is a relatively slow process, but there are real possibilities of bigger impacts. The Arctic is not a significant problem as the ice there is floating on the ocean, so doesn't raise levels by melting. The big disasters-waiting-to-happen are the Antarctic and Greenland, where vast amounts of ice are sitting on land and adding their water content to the oceans would be catastrophic.

Just taking Greenland as an example, there is far more ice there than most of us imagine. Talking about 'the Greenland ice sheet' makes it sound like a thin layer on a fairly insignificant island that won't cause much trouble. But that 'sheet' is mostly over 2 kilometres thick – half as high again as the UK's highest mountain, Ben Nevis. And it covers over 1.3 million square kilometres, the surface area of France and Spain combined. That ice is melting, and the rate of melting is accelerating. It was estimated that in 2012, the annual loss was already four times as much as it had been in 2003.

By 2016, Greenland was losing about 280 billion tonnes of ice a year. That's a huge amount of ice, though at that rate it would still take several hundred years for the ice to totally disappear. However, there is a far worse scenario. Not all the disappearing ice is simply melting and running off to the sea. When there's a crack in the ice, the meltwater can pour through and eat away at the base of a slab of ice, resulting in whole sections of the ice sheet sliding into the water. If the entire Greenland ice sheet ended up in the ocean, sea levels would rise by about 7 metres.

The same concerns apply to the Antarctic, where large chunks can be precipitated. If the entire Antarctic became ice-free, the resultant sea level rise would be 57 metres. Obviously, these are extremes. But even a 5-metre sea level rise would threaten large portions of the UK and US coast, swamping coastal cities such as London and New York, and would have devastating effects in low-lying coastal regions, such as the Netherlands in Europe and Bangladesh in Asia. A storm surge in 1998 left 65 per cent of Bangladesh under water – the more the underlying rise, the more likely that vast swathes of the country could be devastated when storms push waters even higher.

IN THE GREENHOUSE OF LIFE

It might seem that sea levels have little connection to energy consumption – but human-generated climate change is caused by the way that we generate our energy. Energy generation, as we'll see in more detail in a moment, produces greenhouse gases, which have a sometimes-useful property of letting energy from the Sun in, but preventing it from escaping back out again, hence acting a little like a greenhouse. The most widely publicised of these gases is carbon dioxide (CO_2).

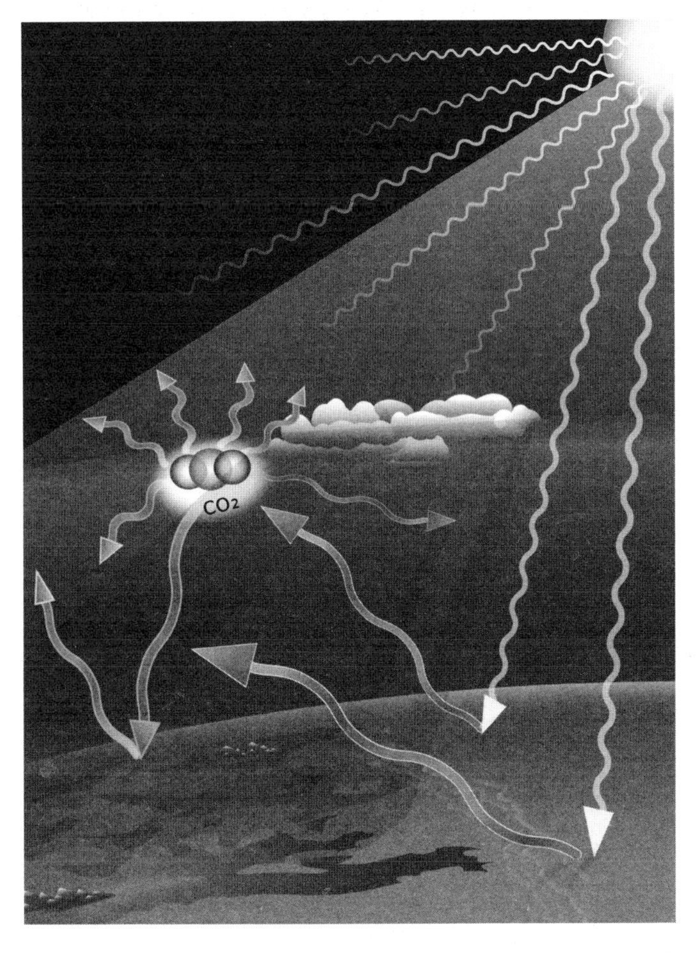

Infra-red radiation from the Earth is absorbed and re-emitted by greenhouse gas molecules such as CO_2, partly back towards Earth.

It's worth stressing that atmospheric carbon dioxide is not a bad thing. The gas is produced in the combustion of carbon-based chemicals, when we burn coal or oil or gas, or when we eat food. CO_2 is an essential for a survivable world for two reasons. One is that plants – at the heart of the world's food cycles – depend on it. As we've seen, plants get their energy from the Sun, but they need carbon to construct new cells to be able to grow. That carbon comes from the carbon dioxide in the atmosphere. Extremely neatly, what is effectively a waste product for us is an essential nutrient for them. What's more, in extracting the carbon, the plants release oxygen – again, an essential for us animals to survive. It's mutual benefit on a vast scale.

The other essential positive role that carbon dioxide has is, strangely, *as a greenhouse gas*. Without its greenhouse gas action, Earth would be largely uninhabitable. The greenhouse effect means that the Earth's average temperature is brought up from a chilly –18°C (0°F) to a balmy 15°C (59°F). That's the difference between life and death. At the temperature we would have without the greenhouse effect, liquid water would be rare and life probably non-existent.

Having some greenhouse gases, then, is great both for life in general and for human beings in particular. But levels have been rising since the 19th century, and that growth is accelerating alarmingly. It's not just carbon dioxide, either. Water vapour and methane, for example, are both more powerful greenhouse gases than carbon dioxide, causing more warming from the same amount of gas. Methane is around 23 times as powerful a greenhouse gas as carbon dioxide and is both produced by agriculture and emitted from permafrost. There are vast amounts of methane locked away in frozen regions, but this is in danger of being released more and more as average temperatures rise, producing a positive feedback effect.

Another agricultural greenhouse impact is from nitrous oxide. This is a particularly pernicious greenhouse gas as it stays in the atmosphere far longer than carbon dioxide. Over a hundred-year period, one tonne of nitrous oxide will make around 300 times the contribution to climate change as a tonne of carbon dioxide. Nitrous oxide release can be a particular problem when excess fertiliser is used.

Agriculture contributes in total around one-fifth of the global greenhouse gas emissions. As well as fertiliser, a significant proportion of this is the methane produced by sheep and particularly cows. (The methane is often blamed on cow farts, but ruminants primarily burp up methane produced by bacteria in their stomach.) We can reduce the impact of this by cutting down on the amount of meat we eat, particularly beef. However, it is possible to almost entirely remove bovine methane production.

There are two approaches, both taking on the bacteria that live in a cow's gut. Unlike cows, kangaroos emit very little methane this way. While switching to kangaroo meat would be one way to do something about this, a more acceptable way might be to transplant non-emitting kangaroo gut bacteria into cows. Even easier, though, would be to make use of the red marine algae *Asparagopsis taxiformis*.

This organism stores up a naturally produced chemical called bromoform. It has been shown that adding just a percent or two of red algae to a cow's diet can bring down the cow's methane emissions by around 99 per cent. The bromoform seems to disrupt the bacteria enzymes that produce methane during digestion. And there would be a double benefit. As well as cutting out most bovine greenhouse gas emissions, it has been estimated that around 15 per cent of cattle's diet is wasted on methane production. It seems that taking this step would also bring down the energy consumption (and cost) involved in rearing a cow.

YOUR GREENHOUSE GAS EMISSIONS

To get a feel for the impact on greenhouse gas emission of changing our diet (or cows' diets), it's worth getting a measure of how much greenhouse gas you as an individual are responsible for being produced (directly and indirectly) each year. After all, the whole purpose of this exercise is to see what makes you *you* – and in the modern world, your energy usage reflects what you do from day to day. The figures are typically given in 'tonnes of carbon dioxide equivalents'. That 'equivalents' part reflects what we've already seen that CO_2 is not the only greenhouse gas, so other contributors such as methane are converted into the equivalent amount of carbon dioxide to allow the whole thing to be added up.

The average UK citizen generates around 7.4 tonnes of CO_2 equivalents a year. The amount varies significantly from country to country, depending both on the typical consumption of an inhabitant and on the way that that country generates its electricity. So, for example, the average US citizen generates around 20.2 tonnes, the average Australian 22.8 tonnes (Australia has particularly coal-heavy electricity generation), the average Canadian 19.4 tonnes, Germans 11.1 tonnes,* Swedes 5.4 tonnes and Indians 1.8 tonnes. To put our dietary contribution into context, if a typical meat eater switched to a vegetarian diet, they would reduce their contribution of greenhouse gases by around 0.8 tonnes a year.

Transport is a way that many of us use energy to make a significantly larger contribution to climate change than our diets. The average contribution of private car use is in the region of 2.4 tonnes a year. This can be removed by switching to an electric vehicle (though these do involve significantly higher carbon emissions in their production than a conventional car) or reduced by making more use of public transport. If, for example, it were possible to switch all your car use to buses, you would be responsible for around 0.6 tonnes a year.

If you fly regularly, you will be making an even greater contribution of greenhouse gases. Not only does it take a lot of energy to keep you in the sky (and electric planes remain highly unlikely for anything

* The German rate is higher than the UK partly because, after having taken the bizarre decision to shut down their nuclear power plants, which have no carbon emissions, they now rely more on coal.

other than very short hops), but the altitude at which planes emit their greenhouse gases gives the same amount of emissions a significantly larger impact on the climate. As a result, one typical long-haul return flight produces almost as much in emissions as a whole year's car use.

The biggest improvements in emissions from travel come when we can switch to electric vehicles – trains being best of all. However, that assumes that the electricity production itself does not result in greenhouse gas emissions. This is true of wind, solar, wave, geothermal, hydro and nuclear but, of course, not when electricity is produced using coal, gas and oil. These fuels also crop up in heating our homes and in industry and between them are still the biggest factor, particularly in countries that are highly reliant on coal for electricity generation.

There is no doubt that solar panels have a big part to play in changing this. The price of solar energy is coming down all the time as the technology becomes more efficient and new variants are introduced. Solar is the most logical energy source, because it involves going straight to the main fount of energy in our solar system, the Sun. With suitable means to store energy (because, let's face it, solar is not a lot of use at night), some countries, such as the USA, could easily provide their entire electricity requirements from solar.

Sadly, this isn't true everywhere. The UK, for example, which can only manage a reasonable solar energy harvest during summer, is unlikely ever to be able to rely entirely on this technology. We can get significantly more from wind and wave – but we are likely to be left with a gap. In principle, this could be filled by importing solar-generated energy from somewhere like North Africa which has a huge solar potential and plenty of unused land – but there are serious political risks to being dependent on imported energy. This means in the short to medium term, despite its negatives, nuclear remains one of the best options to reduce our dependence on fuels that generate greenhouse gases.

The alternative is to put more effort into removing emissions from fossil fuels by using carbon capture and storage. Carbon dioxide can be taken out of the emissions from power stations (or even out of the air) and stored away deep underground. This is already feasible, but currently is expensive and needs more research and development to make it commercially viable.

KEEPING ENERGY FOR A RAINY DAY

One other huge proviso that we skipped over rather quickly above was 'with suitable means to store energy'. This is one the biggest areas of research required to enable us to move away from greenhouse gas emissions. Better, cheaper batteries are essential both to make electric vehicles mainstream and to be able to balance out the output from variable sources such as solar and wind. There are other ways to store energy – for example, the Dinorwig power station in North Wales pumps water up to a high-level reservoir, from where it can flow down and power turbines when demand is at a peak. Stations like Dinorwig store gravitational rather than chemical potential energy. Gravitational stores are only a limited answer, though.

We may need more gravitational stores, but the dams required cause environmental concerns and as battery technology improves, smaller, more local battery-based stores are likely to be more attractive. One other alternative is hydrogen. Hydrogen is a great fuel from a climate change viewpoint, because when it is burned it produces only water – no CO_2 – for the simple reason that there's no carbon in the fuel. At one point there was a lot of excitement about hydrogen as a fuel for cars, but for that use it has significant limitations.

Firstly, hydrogen isn't the safest of materials to be put in the hands of the general public. It is explosively flammable (as disasters with hydrogen-based airships like the Hindenburg demonstrated). And setting up a hydrogen fuel network supplying nationwide filling stations would be extremely expensive. (At the time of writing, there are only a handful of hydrogen filling stations in the UK.) However, none of these issues apply if hydrogen is used as an energy storage medium for power stations. Spare capacity could be used to split water into hydrogen and oxygen, the hydrogen would be stored and then used to generate electricity when required. Hydrogen storage is likely to play a part in any future energy network.

It's also worth stressing that even if we could magically switch all our electricity production overnight to green generation, we would not have done away with our use of fossil fuels. Leaving aside transport, in the UK, for example, 83 per cent of homes are currently heated using gas. That's around 23 million properties. Others use oil. And there's heavy industrial use of fossil fuels too. This is one of the reasons that those taking part in 2019's Extinction Rebellion demonstrations demanding that the UK decarbonise in six years demonstrated worrying ignorance. It will take far longer to replace all those gas heating systems, and at the moment there is little incentive for homeowners to do so. Apart from making the change financially viable, we also need to be generating enough green electricity and to have better electrical heating systems available.

One remaining concern in terms of your personal energy consumption and its impact on the environment is where two of these factors combine. We know that agriculture is responsible for a sizeable chunk of greenhouse gas emissions, as is transport – but what about the transport of food? The concept of 'food miles' – the distance food travels to reach your plate – is often used to emphasise the benefits of eating locally produced food. But it's a concept that is nowhere near as straightforward as it seems.

Often, locally produced food is fresher and more pleasant to eat. Which means it's a good thing. But just because food has been shipped from elsewhere does not necessarily mean that it results in worse carbon emissions. Flying food from a distant country is always a bad move, but shipping has a relatively low carbon impact, and can make less of a contribution than where, for example, heated greenhouses are needed to grow tropical produce in temperate climates. And driving just a few miles to pick up farm-fresh food will produce a similar amount of emissions to walking to a supermarket to buy something that travelled the length of the country.

The energy that you consume, then, has an impact not just on your life, but the larger world. At the time of writing, there have been wide-ranging demonstrations in Europe primarily by younger people concerned about climate change. They rightly highlight this as one of the biggest threats facing humanity. But they do us no favours by framing the threat in apocalyptic terms, demanding changes that are impossible to achieve. We need to bear in mind how integral our energy use is to our lives – there is no point demanding, as some of these demonstrators have, that we magically transform to a totally different society overnight. The changes required to deal with climate change are massive and will take time.

That being the case, it's important that we put effort not only into reducing our greenhouse gas emissions but into mitigating the impact of climate change and into removing greenhouse gases from the atmosphere. These areas fall outside our mission here to explore the individual energy needs required to enable you to function, but are hugely important to the bigger picture.

For now, though, with your energy, food and water requirements satisfied, it's time to move on to another essential consideration. How your species, *Homo sapiens*, got here in the first place.

6

A DIFFERENT APE

When my children were at secondary school, one of their classes produced a poster showing gorillas, suggesting we really should look after these wonderful animals because they are our ancestors – because we evolved from them. In reality, we didn't – but as we have seen, we did evolve from a common ancestor with the other great apes. By looking back in the fossil record, we should be able to deduce something about what makes you what you are today.

GETTING STARTED

Before we take a look at your fossil predecessors, though, it's worth thinking briefly about where they came from and a remarkable natural disaster that could be considered responsible for making it possible for you to be here now.

We've looked into the origins of life and complex organisms. Over many millions of years, evolution has produced more and more variations on the theme of life, leading about 540 million years ago* to the development of the first animals with backbones – the vertebrates of

* All the timescale figures in this section are approximate, but give a feel for the period of time involved.

which you are an example. Initially sea creatures, vertebrates first started emerging from the water, if only briefly, around 500 million years ago.

At first, they didn't find too much of interest on the land – plants only made it ashore about 465 million years ago. A little less than 400 million years ago we see the land-visiting water creatures developing into a four-legged form – the predecessor of many familiar species, including birds (which, as direct descendants of dinosaurs are technically four-legged tetrapods, even though two of their legs are now wings). Probably around 200 million years ago, the first mammals developed and would live alongside and be dominated by the increasingly powerful dinosaurs. And then disaster struck. If you happened to be a dinosaur.

A rocky object around 10 kilometres across came hurtling in from space, smashing into what's now the Gulf of Mexico at around 20 kilometres per second (12.5 miles per second). It produced the Chicxulub crater, a vast scar on the Earth around 200 kilometres (125 miles) across. The result of the impact would have been devastating, releasing around 5 billion times the energy produced by one of the nuclear weapons dropped during the Second World War. The initial result would have been a blast of heat and shock that produced widespread devastation, followed by several years of cooling from the vast amount of dust thrown up into the atmosphere. The resultant climate change finished off the large dinosaurs and transformed the environment for the surviving mammal population.

If it hadn't been for that strike, it's not obvious that mammals would ever have achieved their dominance – without which it's very unlikely *Homo sapiens* (and hence you) would have developed. Without that crisis, there's no reason why dinosaurs and their descendants would not have continued to thrive and dominate. Mammals would have been likely to have remained niche, small animals. Think of the position that reptiles occupy in the world today – not trivial, but equally hardly dominant. Without doubt, part of what makes you *you* is the arrival of the Chicxulub impactor, 65 to 66 million years ago.

THE BROKEN CHAIN

Every few years, fossils of a new form of early human, or potential human ancestor will be discovered and almost inevitably we will be treated, even in the relatively sober parts of the media, to the term 'missing link'. This concept implies a clear, unbroken chain, linking us back through various earlier species to the very beginning of life on Earth. In one sense this is not a totally ridiculous picture. As we saw in Chapter 2, there are indeed chains of descent linking you to your ancestors and if we had perfect information, we could indeed follow these all the way back from you to the first eukaryote and back before that to the origin of life on Earth. But this is quite different from the chain that tends to accompany the missing link concept, which portrays a clear evolutionary process that has brought humanity to greater things than any other animal.

This chain has often been represented by an image that is both as familiar and as inaccurate as the graphic of an atom showing planetary electrons orbiting the atomic sun of the nucleus. In the image, a series of increasingly human-like ancestors leads us from something like a modern ape to a human. And the 'missing link' here is conceived as one of these steps in our progress that we don't yet have evidence for.

Evolutionary chain of man.

However, as we shall discover, such an apparently directed chain is entirely imaginary. The central problem with this viewpoint is that it makes evolution something more than it really is. To get the picture right, we need to take a quick step back and to make sure we understand the implications of evolution. As we have seen, if you break evolution down to its basics, it is impossible to argue against it – however, when it comes to applying it, particularly applying it to humans, all kinds of objections bubble up.

At the heart of evolution is the idea that if you have a way of passing on characteristics to descendants, then those with characteristics that make them better able to survive in the current environment are more like to pass on those characteristics than those who don't have them. Take a silly example. Let's imagine there's an animal called a slugoid. Half the world's slugoids are water soluble and half aren't. After a few days of heavy rain, almost all the slugoids still alive will *not* be water soluble. Only the slugoids that survived will be able to have offspring – so before long the entire population of slugoids will be waterproof. The species will have evolved.

No one can sensibly argue with evolution at this level. As we have seen, though, some feel that this is only a mechanism for micro-evolution. It produced a slugoid better able to survive the rain – but it's still a slugoid. It won't turn one into a weasel. For these objectors to evolution, there isn't a way for one species to change into another. But as we saw in Chapter 2 with the rainbow analogy, this is a misunderstanding of the nature of a species, which is an arbitrary label, rather than an immutable thing.

Some people object to evolution on religious grounds – but there really is no reason to do so. It's a bit like objecting to gravity on religious grounds. Evolution is just a simple following of logical consequences of the way organisms are put together and replicate. It can be easy to think when some evolutionary biologists are also strident atheists that somehow evolution is incompatible with religion, but if you believe in

God, then evolution is a very sensible way for the deity to make things happen.*

There are some restrictions to this compatibility, mostly connected to timescales. Large-scale evolution – getting from, say, a single-celled organism to you – is a slow process. This isn't true, though, of all evolutionary processes. For example, there's an insect called a peppered moth (*Biston betularia*), which has colouration that makes it an excellent match to the lichen-covered trees it lives on. The initial development of this camouflage may well have been a long-term process over thousands or millions of years, when insects that happened to be better camouflaged were more likely to survive and pass on their look. However, during the industrial revolution the polluted air in some areas killed off the lichen, exposing tree bark that was made dark by soot.

The peppered moth in its light and dark forms.

* Of course, biblical stories, such as that of Noah's ark, are not compatible, but plenty of religious believers are happy to see this as a way of putting across a moral message, rather than true history.

As a result, over just a few generations, moths with darker markings were more likely to survive being spotted by predators than their lighter cousins, and so were able to reproduce. The population of peppered moths got much darker over a few tens of years. Similarly, with clean air legislation, the trees have returned to their lighter lichen and the moths have reverted to being light-coloured again.

Nonetheless, to enable the panoply of evolutionary development will have taken many millions of years. This is because evolution is a very inefficient mechanism. We're used to something like a car being designed. It can be a matter of months to go from the design of last year's model to this year's new look. But if evolution were in charge of automobile design, it would take far longer. This is because rather than, say, changing the shape of a panel which we know will make the car more aerodynamic, evolution would, over time, throw up all kinds of changes of shape, many of them making the car worse.* Eventually, though, the changes that fitted best to the environment would be kept. The chances are it would take thousands of years and very many discarded vehicles to get to a car that is a better fit to requirements.

This is why an objection to evolution from the maverick physicist Fred Hoyle doesn't really work as he thought. As we have seen, Hoyle supported an idea called panspermia which suggested that life, rather than evolving on the Earth, had arrived from space (where it had more time to develop). He argued that expecting life to start spontaneously here was a bit like thinking that a hurricane blowing through a scrapyard might accidentally assemble a jet aircraft. And that would be the picture if we expected evolution to come up with you from nothing in one go. But the reality involves vast numbers of tiny changes – some good, many not – over billions of years.

* Admittedly, if you've seen some car designs, such as the Ford Edsel or Tesla's Cybertruck, this appears to happen in real car design too.

We now know that the Earth is around 4.5 billion years old, and as we've seen, it's thought that life has existed for around 4 billion of those years, giving ample time for wide ranges of species to form. When the Victorians realised the timescales required for evolution to work, there was still a common belief that the Earth had been created around 6,000 years earlier. Most religions accepted that evolution made sense and it was the 6,000-year timescale – which was dependent on some very imaginative calculation from ages in the Bible – that was wrong. A few groups do still stick to this 'young Earth' idea, but there is no theological reason to do so.

The other reason that some with a religious belief have a problem with evolution is the concept of teleology. This concept dates back to pre-scientific thinking. It was common, for example, in Ancient Greek philosophy for the most fundamental cause of something to be considered its purpose – what it was *for*. Science does not consider teleological causes. Science can tell us how and why something happens, but not its purpose. This can feel at odds with a religious belief – but it is perfectly possible to accept that humans are special in your religious context without needing evolution itself to be directed. And just because science doesn't consider purposes in natural processes does not mean that purposes don't exist. We do things with a purpose in mind all the time, as do many other organisms. It's simply that for evolution, purpose cannot be a contributory factor.

THE NEVER-ENDING STORY

Evolution has shaped us in response to our changing environment and it is easy to forget that our evolution has not gone away because humans have arrived – once again this is a teleological error. We are still evolving. Everyone's DNA contains small mutations which may, randomly cause a change that could be beneficial for a particular environment and kept. This has certainly happened during the existence of *Homo sapiens*

– those small differences in appearance we use to come up with racial labels are one example – and it continues to happen.

One simple example is lactose intolerance. Originally, *Homo sapiens* was no different from the rest of the mammals. Our ancestors would have drunk milk while infants, but after they were weaned, they would have lost the ability to digest it. However, around 9,000 years ago, in some parts of the world, a mutation produced a failure of the mechanism that switched off the gene producing the enzyme that enables infants to digest lactose. Those of us who have this mutation – far more common, for example, in those of European descent than in those from other continents – continue to digest and enjoy milk and milk products into adulthood. Those without the change, around two-thirds of the world's population, don't.

There is, however, one big difference to the way evolution is acting on us now. As we will discover in more detail in Chapter 8, we modify our environment in ways that have never been possible in the past, changing our nature without an evolutionary component. So, for example, while some organisms evolved the ability to fly, we haven't – but we can now fly thanks to technology. This is a hugely important aspect of what it is to be human, but it is a mistake to think that this means we are not still evolving.

Our technological capabilities have no influence on the existence of evolution. Instead, we are changing the environmental circumstances that shape our evolution – and so the evolutionary outcomes will also be altered. Some worry that the result will be a weakening of the species. Evolution works because some genetic variants support survival better than others. But if we use our technology to enable those who would not otherwise have survived to produce offspring, the critics would argue, then we are not allowing evolution to take its course and remove these 'weaker' genes.

Such a viewpoint is mistaken. It misses out on both the random nature of evolutionary change and the fact that it is entirely possible for factors that reduce survivability to accompany factors that we regard as positive. It certainly may be the case that some evolutionary changes resulting from our modification of the environment will not be beneficial – but equally some could be positive, and we also need to add in the many advantages that our science and technology bring us.

NO GUIDING HAND

The important thing to bear in mind when thinking about our evolution is that the process doesn't make the surviving organisms in any sense *better* overall. Unlike the chain of man image, evolution doesn't have a direction in which it aims to gradually make a more and more wonderful human being, ending up with you. Evolution has no sense of the future – it doesn't know where things are going. It is just a response to the here and now. This means that there are often evolutionary dead ends and that evolution can move a species away from what we consider to be more 'advanced' features, because they get in the way of the survival of that species.

Biologists sometimes go a bit far with this point and say there's nothing special about humans. There is a lot that is special about humans, primarily because of our ability to go beyond evolution, to use our creativity to enhance ourselves and intentionally modify our environment, rather than be modified by it. But what is certainly true is that *Homo sapiens* is not a pinnacle of evolution – you are just a step in a process that has more in common with a drunkard's random, undirected walk than with a march towards perfection. You are not special in evolutionary terms, even though you *are* special in your ability to go beyond the capabilities provided by evolution.

This means that there is no inbuilt tendency to some greater goal on any particular measure. It is perfectly possible for humans to evolve, for example, into less intelligent beings. We even have what may be evidence of this happening. Back in 2004, the news was full of the discovery of fossil remains of real-life 'hobbits' – human-like creatures only about one metre tall on the Indonesian island of Flores, reminiscent (in the imagination of journalists and some scientists) of the miniature human-like creatures in the fiction of J.R.R. Tolkien.

The skeletal remains on Flores had clear resemblances to humans in, for example, the shape of their skulls. Yet they were quite different from *Homo sapiens* in other ways, including having particularly small brains. This might not have been surprising if they had been very distant members of the evolutionary tree. But the hobbits of Flores seem to have lived only about 20,000 years ago, when *Homo sapiens* had already been in existence for a good 180,000 years.

The fact that *Homo sapiens* was around at the same time as the newly discovered *Homo floresiensis*[*] in itself shouldn't have been too much of a surprise. After all, in Europe, humans lived alongside the species *Homo neanderthalensis* – Neanderthals – until around 27,000 years ago, a coexistence that has been graphically confirmed by the discovery of Neanderthal genes in human DNA, particularly that of Europeans, implying a degree of interbreeding. But the implication that *Homo floresiensis* had evolved from larger, probably more intelligent hominins[†] challenges that misleading feeling that evolution should involve an inevitable upward climb of progress.

It's not uncommon on isolated islands, perhaps due to limited resources, for large animals to evolve into smaller versions. Many

---

* Apparently, the discoverers of *Homo floresiensis* originally wanted to call their find 'Homo florianus', until it was pointed out that instead of meaning 'of Flores', this meant 'flowery anus.'

† When dealing with different ape species, some extinct, it can be difficult to know how to apply collective labels. 'Hominids' generally applies to all the great apes, so includes gorillas, chimpanzees, bonobos and orang-utans as well as *Homo sapiens* and our collective extinct relatives. 'Hominins', according to some, excludes the other great apes, only taking in everything after we split off from the chimpanzees, our closest living relatives – though others bundle chimpanzees and their predecessors in as well. And 'humans' only refers to *Homo sapiens*.

islands, for example, were inhabited in prehistoric times by miniature elephants (surely the scientists who talk of bringing back mammoths are missing a trick, as miniature elephants would be much more popular). It should not be surprising, then, if an earlier, larger hominin species evolved into hobbits.

GETTING FROM THERE TO HERE

If we are to successfully get a feel for the contribution that your evolutionary predecessors made to making you *you*, we need to look back in time. However, when trying to piece together what the fossil record tells us about the predecessors to humans, and how we are related, we have a truly messy task. Palaeontologist Henry Gee provides a powerful illustration. Just imagine that we could follow along the various evolutionary branches that lead up to you. We might have a picture of how you were related to all the other hominin developments from our common ancestor with chimpanzees that looked something like the diagram below, where your ancestry is the heavier line, starting in the distant past at the left and coming up to the present day at the right:

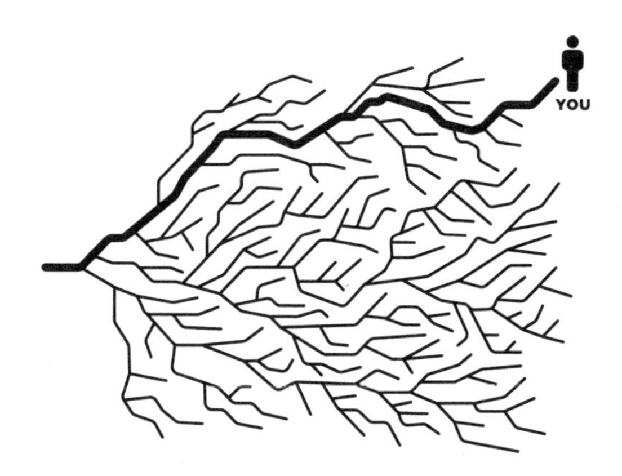

Hypothetical hominin tree, based on illustration by Henry Gee, reproduced with permission.

Unfortunately, we don't have information that is anywhere near as accurate as this. Although this diagram is a best guess, it is only a guess. The reason we can't produce such a diagram for real is that the fossil record is staggeringly incomplete. We know that there are bits missing, of course. But the reality is that what we have is more like a missing *tree*. All that we have are some scattered bits of twig, with reasonable data on where these bits fit on the time axis, but with very limited certainty on where to put them vertically to show relationships. The reality looks rather more like this:

Hominin fossil record, based on illustration by Henry Gee, reproduced with permission.

What we are seeing here is an ancestral join-the-dots puzzle. It certainly could be used to produce the route through to you shown on the previous page, but there is a whole lot of speculation involved in coming up with that particular set of ancestors. There are plenty more possible routes. For example, take the three options on page 99.

While the third, wildly zig-zagging option feels unlikely, it's a salutary reminder that evolution is not in the business of aiming for a particular goal. We arrive with you as a result of a whole collection of accidents and of changing environments that make different developments possible.

Example paths through the fossil record, based on illustration by Henry Gee, reproduced with permission.

The reason that it is so difficult to build a clearer picture is that fossilisation is an exceptional process. Although there are a number of ways that fossils can form, a common cause is that the organism becomes covered in mineral-rich water which fills any cavities and deposits minerals, preserving the structure. In the vast majority of cases, a dead animal or plant will simply decay, leaving no structure behind. Hence that very sketchy set of remnants of the actual tree.

To make the attempt to discover your fossil past even harder, land animals are significantly less likely to form fossils than sea creatures, for the simple reason that they are less likely to end up in mineral-rich water after death. If you are going to be fossilised as a land animal, though, our predecessor hominins' habit of living near water at least gave a chance of being preserved. This compares well with the ancestors of the other modern great apes, which tended to live in tropical forests or at high altitude, where they were more likely to decay than be preserved. Their fossil record is significantly worse than our own.

Where there is DNA available, we can be clearer. All living things use this complex molecule (we'll explore the way it works in more detail in Chapter 9) to pass on information from generation to generation. And all living things have some shared parts in their DNA information, giving us a strong indicator that they came from a common source. Where we have DNA from, for example, Neanderthal remains, we can use that to see how we differ from *Homo neanderthalensis* and to get a feel for when the branch in the tree that separated the two species occurred. However, the further back we go in time, the harder this approach becomes.

Over time DNA deteriorates. It's a complex, distinctly fragile substance with a half-life of about 520 years. This 'half-life' term means that after 520 years, around half of a sample of DNA will have decayed into useless chemical fragments. After another 520 a half of what had remained is lost. The absolute limit is thought to be around 1.5 million years, beyond which no usable DNA will remain. Apart from meaning that the *Jurassic Park* idea of harvesting dinosaur DNA from bloodsucking bugs trapped in amber fails (such DNA would be over 60 million years old), it also means that finding usable DNA from early pre-human hominins is often not possible.

THE IMAGE THAT WON'T GO AWAY

Despite all efforts, it's difficult to get that image of the chain of human development out of our minds. Historically, this has meant that palaeontologists have often assumed a simple, linear step-by-step ordering of fossils, as if that branching tree that we saw on page 97 provided a clear route map. This went along with an assumption that the progression of pre-human development went from one hominin to the next chronologically, rather than there being several different species around simultaneously. Because we are the only species of hominin alive at the moment, it's easy to assume that there was always only one species at a time – but in reality, a single living species has been the exception rather than the rule.

It was because of this linear viewpoint that the most famous archaeological hoax in history was so successful. In 1912, a British amateur archaeologist named Charles Dawson claimed to have discovered what was inevitably described as the missing link between humans and the apes. Dawson's story was that a worker from a gravel pit at the village of Piltdown, near Uckfield in East Sussex, had alerted him to the discovery of a small part of a skull. Dawson brought in Arthur Smith Woodward, an expert from London's Natural History Museum: together they investigated and uncovered more of the skull of what would become known as a new species, *Eoanthropus Dawsoni*.

The fragments that the pair found were reassembled to form a skull unlike any other previously known. Its upper parts were not unlike a human, though with a relatively small brain, but its jaw was far more like that of a chimpanzee. This fitted wonderfully with the narrative of the 'chain of man'. The find, however, was immediately controversial.

Some experts were concerned that the teeth alleged to be part of the find were incompatible with the rest. Nonetheless, for many Piltdown Man was the real deal.

By the 1950s, though, it had been conclusively proved that the skull was a fake, combining what some early observers had suggested: a small human skull assembled with an orangutan jawbone which housed filed-down chimpanzee teeth. The bones had been stained to give them a uniform and aged look. Where some earlier doubters had suspected a simple error in putting together bones that were from the same site but didn't belong together, these modifications made it clear that it was an intentional hoax.

A wide range of perpetrators for the Piltdown Man have been put forward over the years. Most bizarrely, it was once suggested that the Sherlock Holmes author Sir Arthur Conan Doyle was behind the hoax. Doyle would have known Dawson as they were fellow members of the Sussex Archaeological Society, and it has been pointed out that Doyle referred to faking bones in his novel *The Lost World*, published coincidentally in 1912. Fun though such speculation is, it seems far more likely that Dawson was behind the fake. Not only did he have the most to gain from the publicity, he was later discovered to have a whole range of fakes in his collection. Whoever was behind Piltdown, though, it shows how easy it is to construct models of prehistory from fossils that are more about supporting our preconceptions than about science.

One of the reasons that the Piltdown hoax was taken seriously by some was that it supported a theory that walking upright was part of a parcel of developments that followed the evolution of large brains. But as more fossil finds were made in Africa and the far East, it became clear that there were hominins that still had relatively small brains that were already walking upright, most notably in the cases of 'Java man' and 'Peking man' which proved to be the first discoveries of an earlier species than *Homo sapiens* now known as *Homo erectus*.

Although the fossil record is still very sparse, a combination of more finds and DNA evidence has made the 'out of Africa' hypothesis, where hominins developed in Africa and dispersed in a number of waves into Asia and Europe (quite probably also moving back into Africa to add confusion to the timeline) very likely. As we have seen, a small part of our DNA is contained in tiny structures called mitochondria that act as the power sources of cells. This DNA's simplicity makes it particularly useful for tracing potential origins of individuals – as was the case with the mitochondrial Eve calculation (see page 17). A study in 1987 showed the most variation in mitochondrial DNA was found in Africa, while that found elsewhere all appeared to be variants of an ancient African form.

A particular difficulty comes in assigning fossils to species, especially as the remains are often fragmentary, some consisting of little more than a single bone. Bearing in mind that species are relatively arbitrary labels, quite a number of remains have had to be moved between *Homo* and other non-hominin species such as *Australopithecus* as further data came to light.

This makes things a little frustrating when pinning down where you come from as a species. We know that *Homo sapiens* has been around for at least 200,000 years. As we have seen, there have been other species of *Homo* coexisting with *Homo sapiens* with sufficient sharing of genes to give, for instance, Neanderthals a small part in modern European human ancestry, just as another early human species known as Denisovans, first found in Russia, have contributed small amounts of DNA to the inhabitants of Melanesia, Papuans and Aboriginal Australians.

It's possible to estimate from the rate at which changes occur in DNA when different species split from a common ancestor. From this approach, it has been suggested that chimpanzees and hominins went their separate ways around 5–7 million years ago. We can place plenty of remains in the gap between that distant past and the development of *Homo sapiens* around 200,000 years ago – though as always, we can't say if there's a direct line of descent from the unidentified first hominin to us. For example, a widely discovered early hominin was *Australopithecus africanus*, which dates back around 3 million years – at one time it would probably have been seen as a link to humans. Now we can only say that it was one of many species in that timeframe.

Perhaps the best-known of the really old specimens is a partial skeleton known as Lucy,* found in Ethiopia and the only known example of the species *Australopithecus afarensis* (not because she came from afar, but from the Afar region), dated to around 3.6 million years ago, while a few fragments, also from Ethiopia, have been found of a circa 4.4 million-year-old hominin, *Ardipithecus ramidus* (known as Ardi). By the time we get to this distance in the past, combined with the typically limited amount of remains discovered, it can be borderline whether what's being dealt with is hominin or from the wider possibilities of hominids – with more resemblance to apes and monkeys there is inevitably a possibility that the species is out of a different developmental tree.

Despite this uncertainty, you will regularly see claims of a new human 'ancestor'. For example, in August 2019, newspapers ran a story about the latest fossil find. The UK's *i* newspaper went with the headline 'Meet MRD, Lucy's relative and mankind's oldest ancestor'. This referred to the discovery of a 3.8 million-year-old cranium in Afar, identified as the species *Australopithecus anamensis*. Strictly speaking the 'Lucy's relative' part of the headline is true, in the sense that all animals are relatives – and MRD† and Lucy were relatively close in timeframe. But there is no evidence whatsoever that MRD was our ancestor – and clearly isn't our oldest ancestor, as that would require going back 4 billion years or so and would be something like a bacterium.

* Apparently so named because the Beatles track 'Lucy in the Sky with Diamonds' was popular with the discovery team.

† The 'MRD' name is just a shortening of the specimen's collection identification number.

The same article claimed: 'scientists ... believe that the skull is that of an ancestor of the renowned fossil hominid known as Lucy.' MRD is a relative of Lucy's, certainly, but there is no evidence of MRD being an ancestor. It's worth bearing this in mind when you see such stories: the media, and even some scientists, are likely to claim significantly more than we can actually discover from these remains.

BUT WHY ARE WE LIKE THIS?

It can be quite frustrating to realise how little the fossil record can tell us about how you got to be the particular organism you are. However, this hasn't stopped theorists attempting to explain how we ended up as relatively hairless apes that walk upright and have unusually big and complex brains for our body size.

Some of the theories that attempt to explain the particular form *Homo sapiens* takes feel a little unlikely. For example, in the 1960s, English marine biologist Alister Hardy came up with a hypothesis that a major factor in our evolution was that we developed from an ape that was partially ocean-dwelling – known as the aquatic ape hypothesis. This was picked up and built on by Welsh author Elaine Morgan, suggesting that features such as hairlessness, our levels of subcutaneous fat and our fondness for shellfish were evidence of this origin. Even walking upright was proposed as a benefit for an ancestor which spent a considerable amount of time wading through water looking for food.

The theory is generally dismissed by academics because of lack of evidence to back up what was little more than a collection of features that happen not to be incompatible with such a lifestyle. For a long time, the most commonly supported hypothesis was that at least some of our developments were due to the move from the trees to the savannah as the African rainforests died out. It was suggested that standing upright gave us a better ability to scan the open landscape and freed up hands for using tools or weapons. A lack of body hair was supposed to make us better able to cope with the hotter conditions we faced. But the arguments here are also flawed.

Firstly, early hominins were walking upright significantly earlier than the climate change that made savannah-dwelling more attractive – and, let's face it, there are plenty of other mammals that lived and still live on the savannah, none of which have benefited from a permanent upright stance.* Our two-legged gait doesn't encourage survival either. As any dog owner can testify, four legs gives a significant advantage when it comes to running speed. Our ancestors also lost the large canine teeth typical of great apes and developed thinner skin – both changes offering less protection against predators than their predecessors are likely to have had. And though there are small advantages from freeing up the hands to carry items and use tools, modern chimpanzees seem capable enough of doing this without taking on the disadvantages our upright ancestors faced. Similarly, the heat argument for minimising body hair doesn't seem to have proved a problem for all the other mammals that live in open, hot country.

In reality, it is hard to argue that a relatively hairless, thin-skinned, upright, flat-footed, big-brained ape survived due to any direct evolutionary advantage. Almost all of these developments present a disadvantage in survival terms. This might seem to run counter to evolution, but such a viewpoint takes us back to the teleological myth that evolution has a purpose. We don't evolve in a certain way so that we can benefit from it. We evolve in random ways, and if the overall benefits outweigh the disbenefits, we are likely to keep those features. This can even happen if an evolutionary feature has a negative impact on survival, as long as there are other associated benefits from the change that brought it about that more than counter the negatives.

* Meerkats, for example, do live on the savannah, and regularly stand briefly on their hind legs, but aren't anywhere near being bipeds. The vast majority of savannah-dwellers, both prey and hunters, have never developed this supposedly beneficial posture.

Think for a moment of the peacock. For a prey animal, it's hard to imagine anything more risky, in survival terms, than making yourself highly visible.* However, for the peacock, the sexual selection benefits of their huge display tail – peacocks with fancier tails are more attractive to peahens – outweigh the survival risk from the burden of being so visible to predators and so hampered in motion.

In our case, all those apparent negatives were unlikely to make us more attractive to potential mates, but were more likely to be a side effect of some other, overwhelming benefit. We know that some hominins have been bipedal, walking upright, since at least 4 million years ago and just possibly for 2–3 million years more – though, as always, we need to bear in mind that the early upright walkers are not necessarily our ancestors, as such behaviour could have developed more than once. It's possible that bipedal behaviour was influenced by a number of factors, but one hypothesis is particularly interesting.

This came from English zoologist Clive Bromhall: it seems to be less unlikely than the alternatives, although again it is difficult to present evidence to directly support it. Bromhall suggested that the distinctive features of *Homo sapiens* are no surprise to anyone who has studied chimpanzees in the womb. Before it is born, a chimpanzee has almost all of the oddities of the human form (odd when compared to the other great apes, that is). A chimp foetus, as you do, has a forward-facing head, a skeletal structure idea for bipedal motion, flat feet and face, small teeth, is generally relatively hairless apart from the top of the head, has thin skin and a bigger, more rounded skull.

Bromhall suggests that all of these more obvious oddities in *Homo sapiens* (along with several hidden ones, such as peculiarities in our lungs and aortic arch) echo the physique of infant apes. We retain the features of very young members of the species when we mature, a process known

* Unless you mimic the visibility of a poisonous species, for example duplicating the bright colours of wasps or poisonous frogs. Such a feature is called being aposematic.

as neoteny. Neoteny is not unique to humans – it is particularly common in domestic animals, most notably dogs, which resemble wolf cubs in several ways. The only other great ape to have some neotenous development is the bonobo,* also known as the pygmy chimpanzee, which compared with the chimpanzee has a more rounded skull, flatter face, more of a tendency to walk upright and other neotenous details and traits.

All in all, the human characteristics that you have inherited seem to give you an unlikely chance of survival among the ferocious predators of the savannah. As Bromhall dramatically puts it, 'Instead of creating the incredible hulk, evolution had opted for one of the puniest apes of all time.'† What seems to have kept our ancestral apes of the plain alive is living together in larger groups – a classic adaptation of an exposed prey animal. This still applies to the few apes that are ground-dwellers.

By comparison, chimpanzees do not thrive in large groups – typically around five together is optimal. With many more, infighting in the group becomes too destructive. But around five is simply too small a group size to survive by standing up to large predators, where it's not possible to escape into the trees. Bonobos are totally different from chimpanzees, living mostly peaceful lives in large groups with more of a female dominance than male. Bromhall suggests that bonobos – and humans – are able to live together in large protective groups because they are neotenous – because they retain both infant features and behaviour compared with that of most apes.

Although this theory may not provide a total explanation for the way you look, it does seem a very logical reason why a neotenous

* Apparently the bonobo's name is the result of a typo. It was named after Bolobo, a town in the Democratic Republic of the Congo, but the word was incorrectly transcribed.

† By now it should be clear that, taken literally, this is meaningless. Evolution doesn't decide to create an incredible hulk or a puny ape. But we can assume that Bromhall did not mean it literally.

mutation of a pre-human had a better survival chance in the savannah than a more traditional ape. And it was purely coincidental that this gave you and the rest of us other benefits that followed later through an enlarged brain.

BECOMING A MEGA-BRAIN

Arguably the most important contribution to making you what you are today is the transformation of the brain. Biologists who object to 'human exceptionalism' make the case that most of our intellectually-based abilities are also present in other animals. Tool use, for example, is found in everything from chimps to crows. However, no other animal comes close to the sheer scale of ingenuity and creativity displayed by the human brain. Our neotenous appearance and behaviour may be a side effect of improved social interaction, and it could have resulted in bigger brains, but doesn't in itself explain the additional complexity of human brains.

With the usual proviso that we don't know which of the hominin fossil remains were in our direct line of evolution, we can see a major shift in the past 2.5 million years from brains that were similar in size to those of chimpanzees to the current human brain at around four times bigger. Of itself, brain size isn't a direct measure of intelligence and creativity. An elephant's brain, for example, is about three times the size of a human's, while some dolphins, despite being smaller than us in body size, have larger brains. However, in the absence of preserved ancient hominin brain structures to examine, we are restricted to using the increase in brain size as a measure, while also being able to observe the added complexity of the modern human brain when compared to other modern mammals.

It was long argued that our more effective brains were a simple evolutionary trait. Those with an improved thinking capacity were better able to solve the problems that they faced, so more likely to survive. Over time the human brain got bigger and bigger until it reached the magnificent specimen you are using to read this book today. This same argument is why it often used to be the case that people of the future (or spacefaring aliens) were portrayed in early science fiction as having bulging heads to encompass their ever-growing brain capacity. But there are several problems with this argument.

Pop-culture depiction of an alien with bulging head, from *This Island Earth* (1955).

One tongue-in-cheek issue with evolution producing bigger brains is the cheerleader/football star argument. As anyone who was nerdy at school knows, it's not the ones with the big brains that win the attention of the opposite sex, but rather those who have good looks and sporting abilities. The process of sexual selection might suggest that, if anything, big brains were a negative evolutionary pressure that was naturally self-limiting.

More seriously, if the 'more brains equals better survival' argument holds, we might expect that all species would see a gradual but steady increase in brain power – yet relatively speaking, *Homo sapiens* seems to be highly unusual in this respect. Clive Bromhall explains the increase in brain size as nothing more than another example of neoteny in action – infants have larger heads (and brains) in proportion to the size of the body. This may well be true, but given the way that brain size doesn't equate to intelligence or creativity, we still lack a convincing argument for the origin of the additional complexity that enables us to evolve not just in response to our environment but by transforming our environment.

Here things get distinctly speculative. It seems likely that in part the driver was effectively a new way of using the same physical components of the brain, one that allowed the early humans to detach themselves from the flow of the 'now' and ponder 'What if?' – as some have put it in a poetic fashion, the ability to dream while still awake. There is no doubt that there is something here, based on physical changes, which would result in major evolutionary benefit. However, we do not have a clear idea of what it is.

One suggestion that has been made is that our large brains were only supportable because of the move away from forest areas to shores, particularly coastlines, though without spending a lot of time in the water as suggested by the aquatic ape hypothesis. Two important micronutrients for the brain are iodine and a fatty acid called docosahexaenoic acid (DHA). Both of these are found in diets that incorporate algae (including its large form, seaweed), and seafood that feeds on algae. It is argued that without this move, the enlarged brain could have caused significant problems and might not have developed the way it did.

It's clear there is still much to discover about the nature of the human brain. And even if we can get a feel for the physical changes that made human modes of thought possible, one particularly difficult strand in understanding 'you' is pinning down what you would probably think of as your 'essence', the conscious entity that seems to float somewhere in your skull. What, then, is your consciousness?

7

IS YOUR CONSCIOUSNESS AN ILLUSION?

Ask a scientist to name the biggest unanswered questions in science and they might mention dark matter, dark energy, the combination of quantum theory and general relativity in a single 'theory of everything', the origin of life and more. But one topic that is certain to come up is the nature of consciousness. This inner awareness and mode of thought is (probably) not entirely limited to humans, but it seems far more significant in making us what we are than is the case for any other organism.

DO I THINK AT ALL?

Consciousness is a relatively modern concept. The word 'conscious' comes from the Latin *conscius* which mixes meanings of sharing knowledge, being inwardly aware and feeling guilty. The term 'consciousness' seems to have been first used in the 17th century around the time that French philosopher René Descartes came up with his famous phrase *Je pense, donc je suis* (I think, therefore I am). Obviously people were aware of what we now refer to as consciousness prior to this, but it seems not to have been thought about in the same way, perhaps in part influenced by the greater stress that was put on the soul as a separate entity.

It's easy when thinking about consciousness to get into a tangle – after all, the very action of thinking about consciousness requires … consciousness. When I was young I remember cartoon strips in which there were a whole collection of little people inside a person's head, pulling on various controls to make the robot-like human body function, something reflected to a degree in Pixar's 2015 film *Inside Out*, where several emotions were personified as creatures within the main character's 'mind headquarters'. Of course, any attempt to apply logic to such fictions immediately descends into an endless spiral, as presumably each of the conscious personifications themselves would have their own inner controllers and so on.*

This image of having controllers inside your head, watching the world on a sort of screen at the front and sending commands to the different parts of your body – a model sometimes called a 'Cartesian theatre' (presumably in the US sense of a movie theatre) – is clearly ludicrous, but it's easy to see where the idea originates. Because of the position of your eyes in your head, you see the world as if your conscious 'me' is located somewhere just behind them, looking out on everything around you. Take a moment to try to extrapolate back from the page into your consciousness as you read these words. Where is the 'you' that is reading them? It feels to be roughly between your eyes and back a bit.

Your consciousness is the aspect of your being that makes sense of your sensory inputs – and that's part of the problem, because physics tells us we're not at all good at perceiving what is really there. The philosopher Immanuel Kant† made the observation that although there

* This brings to mind the mathematician Augustus de Morgan's poem Siphonaptera: 'Big fleas have little fleas upon their backs to bite 'em, / And little fleas have lesser fleas, and so, ad infinitum. / And the great fleas, themselves, in turn, have greater fleas to go on; / While these again have greater still, and greater still, and so on.'

† This is the second philosopher already in this chapter, so apologies if you feel that philosophy and science are very different things, and philosophy has no place intruding into a book like this. Unfortunately, it's pretty much impossible to think about consciousness without bringing philosophy in, and, to be fair, this pair both

is (presumably) a reality out there – what he called the *Ding an sich* (roughly, the thing itself) – we can never directly experience it. All we can ever know is what our fallible senses tell us about it. According to Kant, we experience phenomena, but the 'real world' things that cause these phenomena are not accessible. This understanding of reality was, if anything magnified by quantum physics.

THE STRANGENESS OF QUANTUM REALITY

From the 1930s onwards, it became clear that in the world of the very small – the atoms that make up all matter, the photons in a beam of light – reality is very different from the world we experience through our senses. We don't expect, for example, that objects will pass through walls as if they aren't there. Yet quantum physics tells us that quantum particles can do the equivalent – pass through a barrier that should stop them in their tracks and appear on the other side.* This and other strange aspects of the quantum world have been repeatedly demonstrated. We know that quantum theory is right – but it bears no resemblance to our conscious perception of the world around us.

Aside from the way that you experience the world, there is also an aspect of your perception of consciousness itself that is at odds with the scientific picture. You may be able to locate 'you' between your eyes, but if you are honest, that 'you' feels to be detached from your body, something separate – a concept known as duality. This was pretty well the only approach until the 17th century, when the spread of rationalism

made very scientific contributions to philosophy, rather than indulging in pure navel-gazing.

* We recommend that you do not try to run through a wall as, despite being made up of quantum atoms, your body is not good at passing through other solid objects. This may seem an unnecessary warning, but at the height of the US military's attempts to harness psi phenomena, Major General Albert Stubblebine, head of the US Army Intelligence Corps, tried to do just this. All he succeeded in doing was squashing his nose.

made it more acceptable to consider a human being to be a 'meat machine' with no separate essence or soul or spirit.

As we have seen, the word 'spirit' originally comes from the same root as words applying to breath. The Latin equivalent of the verb 'to breathe' was *spirare*. When we died we were said to expire – the breath left us. For the post-Renaissance rationalists, it was unnecessary to separate our consciousness from the operation of the brain. All that you are, according to that viewpoint, is the electrochemical* working of your brain – you have no separate spirit. This is the standard scientific viewpoint now, though it is almost certainly true that the majority of the world's population, either through mainstream religious beliefs or New Age concepts, continues to take a dualist viewpoint.

What science can't do is say that the dualist view is incorrect. It can't be proved that there is no separate spirit. It is perfectly possible to hold such a view within a scientific framework. But scientists would usually argue that there is no need for the additional complexity of a soul, nor is there any known mechanism by which a non-physical entity could be tightly tied to the physical reality of the brain. This implies that the 'meat machine' approach is the simplest concept, which, lacking any other evidence, makes it the one that science adopts. Of course, science also tells us that the world is rarely as simple as we think – and believers would say that their feelings tell them that being human is something more than just the electrochemical activity of the brain. You pays your money and you takes your choice. But it's not an easy thing to sort out – in fact, the whole mind/body issue is often known as the 'hard problem' in psychology.

* The word electrochemical comes up a lot when describing the operation of the brain. While the brain has a kind of electrical mechanism, it's very different from, say, the hard-wired electrical configuration of a computer. In biology, electrical signals are carried by charged chemical ions moving from place to place. One implication of this is that the processes are far more sluggish than in a traditional electrical circuit – but the brain is so complex in structure that it more than compensates for this relative slowness.

THE BRAIN IS WHERE YOU LIVE

What we do know is that your brain has a primary role in supporting your consciousness (whatever that is). Historically, there was some dispute over what the brain was for. Although observations of the impact of brain injury had made the brain a likely candidate for the intelligent bit of the body in Ancient Egypt as far back as 3,700 years ago, and some Greek philosophers supported this idea, others, notably Aristotle, thought that the heart was where 'you' were located (hence all the lovey-dovey heart business on Valentine's Day – don't blame the greetings card companies, blame Aristotle). Aristotle and his followers relegated the brain to an important, but secondary role of acting as the equivalent of a car's radiator by cooling the blood in its complex folds and wiggles.*

By the 17th century, though, when the English physician William Harvey had built on the work of early writers to give a description of the circulation of the blood around the body, the heart was definitively seen to be responsible for pumping that blood, while the brain achieved its rightful place without dispute. (It's hard to understand quite why Aristotle took an opposing view, given the clear mental picture of 'you' being behind your eyes. Perhaps it was because brain damage, devastating though it can be, is less certain to kill you than is heart damage.)

The human brain typically weighs about 1.3 kilograms (3 lbs). Despite only being around 2 per cent of your body weight, the brain uses up to 20 per cent of your body's energy, which is why it was so important to justify its value when considering how the predecessors of humans evolved in the previous chapter. Your brain (and, to be fair, everyone else's) has been described as the most complex structure in the known universe. There may be plenty of strangely complex things out there that we don't know about, but most of what we can observe of the universe (stars, for example) is fairly simple in structure, if big in scale

* Technical biological terms.

– and on Earth there is nothing else that comes close to the intricacy of the human brain.

Looking at an actual dissected brain, there clearly is a lot going on in the structure, with the wrinkly grey matter of cerebral cortex on the outside and the various internal parts such as the white matter beneath the grey, the cerebellum (responsible for aspects of motor control, for example) and the hippocampus, named after a seahorse (though to see this, you have to have a lot of imagination) which is thought to be responsible for memory processing. The brain is divided into two almost entirely independent halves, each primarily responsible for the opposite side of the body, these halves being joined by a chunk of white matter called the corpus callosum.

However, impressive though this is, it's the complexity invisible to the naked eye that is really stunning. The brain has a range of functional regions that handle different requirements – for example, image processing or memory retrieval – but on the whole these aren't clearly identified physical modules, but rather logical constructs that can be located in several parts of the brain working together. The basic functional unit behind it all is the neuron. This is an elongated nerve cell, which can have multiple synapses – electrochemical connections to other neurons. It's these connections that pass signals from place to place when stimuli reach sufficient levels, to provide the equivalent function of the processor in a computer. There are a phenomenal number of synapses in the adult human brain – as many as 100 billion neurons can each have numbers of connections reaching into the billions. The resultant set of potential permutations is far bigger than the estimate for the number of atoms in the universe.*

As well as controlling the physical behaviour of the body, your brain is without doubt responsible for the subjective aspect of what makes

* As we have already seen, the estimate of number of atoms in the universe is 10^{80}. The more honest cosmologists admit this has to be a pure guess, as we don't even know how big the universe is.

you *you*. This becomes clear when someone we know suffers from a degenerative brain disease and the person that they used to be can seem to almost entirely disappear. For those who believe in the brain/spirit duality, it's assumed that in some way the brain is an interface to the real you, and it's the interface that's failing with such diseases – but most scientists dispute this for the reasons mentioned above.

In the end, bearing in mind that we are on a quest to discover what makes you *you*, only you know for sure what appears to be going on inside your head (though modern technology, such as fMRI scanners* do enable us to get broad pictures of what the brain is up to). The only thing you can truly be sure of as far as consciousness is concerned is what you observe of your own consciousness. Some have taken this to the extreme in the philosophical stance of solipsism, saying not only is this all you can know, but is all that can be said to exist. Most of us, however, accept the reality of minds and objects beyond ourselves, even though we can't truly know what is going on inside someone else's skull.

GOING BATTY

This hyper-personal aspect of consciousness was underlined in the 1970s by the American philosopher Thomas Nagel when he decided that we could never really know what it's like to be a bat. Nagel argued that consciousness 'occurs at many levels of animal life, though we cannot be sure of its presence in simpler organisms, and it is very difficult to say in general what provides evidence for it.' This seems to be an assumption made without good evidence, but it does appear reasonable that some other animals have a form of consciousness.

Starting from that assumption, Nagel suggested that if an organism is conscious, then there is 'something it is like to *be* that organism'.

* Functional Magnetic Resonance Imaging. These remarkable devices temporarily turn the water molecules in your body into tiny radio transmitters.

(Run that statement through your mind a couple of times.) If an animal experiences the nature of being whatever it is – if it has a subjective experience of its world, rather than a series of mechanical responses to its sensory inputs – then it is conscious. The implication is that there is something more than just behaviour in these cases, something that lies beneath it.

Nagel chose bats as an example because they are mammals, so probably more likely to have a sense of what it is like to be a bat than, say, an insect would have a sense of what it is to be an insect. However, a bat has abilities and senses that are dramatically different from our own. The bat 'view' of the world, for example, incorporating sonar, is inevitably going to be radically different from our experience based on sight. Nagel then argued that the best we can do is to envisage what it would be like for *us* to be equipped like a bat – but we have no idea of what it is like for the bat.

Nagel concluded that, although our consciousness is a physical process, produced by electrochemical activity in the brain, it is misleading to label consciousness as simply the outcome of neural activity, because what we call consciousness is emergent, something that can't be described simply by reducing it to individual physical causes, but emerges from the interaction of the components that make it up, forming more than the sum of the parts.

The 'What is it like to be a ...' test can be seen as a way to distinguish the conscious from the not conscious. It seems pretty clear that it is not possible to apply such a conception to an inanimate object like a rock; by contrast, surely we can, say, ascribe some form of consciousness to an intelligent mammal. Exactly where we draw the line is harder to pin down. Bacteria and plants seem highly unlikely to be conscious. Insects – well, quite probably they aren't conscious either. But as the bat example makes clear, we are so far removed from being able to get into another organism's mind that drawing a hard and fast line is pretty much impossible.

Some studies have been designed to try to uncover how much an animal understands what it sees. Notably, animals are shown themselves in a mirror with a spot painted on their heads. The idea is that if the animal, seeing the spot in the mirror, tries to rub it off its head, it realises that the image in the mirror is connected to it. Most animals don't recognise themselves but the great apes – excepting gorillas – do. However, in another experiment, chimpanzees, our closest great ape relatives, who pass this spot test, were given the opportunity to beg for food from a human and would do so even if the human clearly couldn't see them (because, for example, the human had a bucket over his or her head)*, suggesting that even chimpanzees don't have our ability to think themselves into someone else's position.

Not everyone agrees with Nagel's assertion that we can never discover what it is like to be a bat – as with almost everything else they consider, there is no consensus among philosophers – but there seems to be agreement that it's not enough to deal with the processes of the brain. We need some way of addressing subjectivity and of uncovering how the sensory experiences of the conscious mind are detected and processed.†

There is a division of thought over whether consciousness is something that sits on top of our thoughts, sensory experiences and responses, so needs explaining in its own right, or is simply the inherent outcome of the interaction of that thinking, sensing and acting, not existing as something independent.

* Psychologists really do know how to have a good time.

† Some philosophers refer to the 'units' of subjective experience, such as the nature of a colour or of a smell as 'qualia'. I think this term, with its echoes of a physicist's quantum, gives the concept an undeserved scientific flavour. The idea of qualia itself is highly subjective and one that not all philosophers are comfortable with.

ON AUTOMATIC PILOT

With the near-irresistible image of the 'you' between your eyes telling your body what to do, it's hard to totally escape the idea of having a separate consciousness – the thing that's in control and that makes things happen. As I write this book, I can imagine that 'conscious me' is deciding what the next word that I'm going to type is and instructing my fingers to hit the correct keys. That feels right for the process of writing a book. And yet there are a couple of problems with that image.

One is that, in practice, I don't consciously decide what my fingers are going to do on the keyboard. I'm a touch typist. I don't look at the keys when I type, and I don't tell my right forefinger to move to the correct position to hit the L key when I type a word with an L in it. If you were to ask me where the L key is on the keyboard, I honestly can't tell you. This is because when you learn something like touch typing (or driving, or playing the piano, say), you only achieve a reasonable skill level once you move from consciously carrying out actions to having some of the subconscious parts of your brain take over the task. There are bits of the brain well away from consciousness that are very good at the 'doing things' side of life and we use a different kind of memory to store away how to do something once we become trained in it and it becomes semi-automatic.

In his book *Incognito*, American neuroscientist David Eagleman points out an excellent example of the gulf between conscious awareness and what the brain actually *does* that anyone with experience of driving can try out. Imagine that you are driving a car in the left-hand lane of a three-lane motorway. You want to get into the middle lane. What do your hands do with the steering wheel in order to accomplish this? Just close your eyes for a moment and think through the action before you read on.

The chances are extremely high that you imagined turning the steering wheel a bit to the right, waited a few moments as you shifted from lane to lane, then re-centred the wheel. That seems to makes sense. However, I don't suggest you try this next time you are driving, as the result would be a crash. What you actually do is to make that turn of the wheel to the right, but to settle back into driving straight down the middle lane you need to briefly turn the wheel back to the left before centring. Your brain knows to do this – but it's highly likely that your consciousness doesn't.

THE DOUBLE-DEALING BRAIN

The brain cheats a lot. This is particularly obvious when we compare our visual image of the world with what's really happening out there. We tend to think of what we see as being a biological equivalent of a camera. The lens at the front of the eye projects an image onto the sensors in the retina, just as in a camera the lens throws an image onto an electronic sensor. Then the rods and cones in the retina, with their optic nerve connections to the brain, assemble a picture. But just as the camera doesn't actually store an image like a physical photograph, but instead holds a collection of zeros and ones representing the scene, so your brain does not project the view onto some kind of internal screen to produce the nice, clear image you appear to see in front of you.

In reality, the signals from the rods and cones in your eyes, funnelled through the optic nerves, are picked up by a series of modules that do things like separate out shapes, deal with blocks of colour and so on. This explains why what you see is deceptive. One obvious example is the fact that there's a blind spot on your retina where the optic nerve connects to it. But you don't see that gap – the brain fills the image in for you. Similarly, your eyes are regularly darting about in very quick little motions called saccades – but your brain irons out the motion-sickness-inducing jerkiness and provides a totally fake still image.

The disconnect between the imagined way our eye–brain combination works and its real mechanism accounts for the wide range of optical illusions that have been produced. There's a beautiful example in the so-called chessboard illusion,* where two apparently very differently shaded squares are actually identical.

The chessboard illusion. Although it is hard to believe, the square marked A and the square marked B are the exact same shade of grey.
Image by Edward H. Adelson

One of the most frequently experienced of all optical illusions is the Moon illusion. If you've ever taken a photograph of the Moon without using a telephoto lens or a telescope, you will probably have been disappointed that it looks so insignificant. That's not a problem with your camera – it's what the Moon really looks like. Our satellite's true apparent size is about the same as the hole in a piece of punched paper, held at arm's length. But for reasons we don't wholly understand, when you see the Moon, your brain inflates it by several times its actual apparent

* See www.universeinsideyou.com/experiment3.html to see an animated demonstration of this illusion which makes it clearer the two squares are the same shade.

size – an effect that seems to be particularly strong when the Moon is near the horizon.

Another example that you probably encounter every day is the moving picture illusion, where a video consisting of a series of still images appears to show smooth movements. For a long time, this was explained as being due to something called 'persistence of vision'. The idea was that your brain hung onto an image for a fraction of a second, and if the next image was presented quickly enough, then the two would somehow merge into each other. The trouble with this explanation is that it shouldn't produce a clear moving image but a superimposed mess. In reality, the illusion of motion from a series of still pictures happens because your brain modules are handling things like straight lines, shapes and movements and your brain is constructing something that fits with the moving real world that it is familiar with – it's cheating again.

This is not some peculiarity limited to the *visual* systems of the brain – all of our conscious experience is manipulated. When you see the flash of lightning, then a few seconds later hear a crash of thunder, you almost certainly know that these are simultaneous events – thunder is just the sound of that lightning bolt ripping through the air – but that's intellectual knowledge.* They don't feel like the same thing. However, when sound and vision come from something for which we have a more inbuilt assumption of connection – for instance when something we drop hits the floor – we hear the sound and see the cause simultaneously. But we shouldn't. We now know that sound and vision are processed at different speeds in the brain. The signals from the event do not arrive simultaneously, but the construct that is our conscious sensory impression combines them.

* And relatively recent intellectual knowledge – thunder and lightning were long thought of as separate phenomena.

IT'S CONSCIOUSNESS, JIM, BUT NOT AS WE KNOW IT

Given that we know that the brain is playing such sophisticated tricks on us to present the sensory image of the world that we perceive, it makes it more likely that the feeling of a conscious 'you', located somewhere in your head behind your eyes and pulling the metaphorical levers to control your actions, is also an illusion. This is not to say that consciousness does not exist, but that the way we perceive it is at best likely to be deceptive.

We might defend consciousness by saying that while it's certainly true that some activities can be pushed into a pretty much autonomous part of the brain and nervous system, this clearly doesn't apply to decision-making. Surely here our conscious mind is responsible for weighing up options and coming to a rational decision? But there is experimental evidence to show that this is not always the case.

In the 1990s, Canadian neuroscientist Antoine Bechara and colleagues ran an experiment where participants were asked to choose cards from four different piles. Some cards gained a player money, others penalised the player by taking money away. It took around 25 goes before the subjects were typically able to identify which piles were on average beneficial, and which were best avoided. However, by measuring the electrical conductivity of the skin, the scientists were able to spot when the nervous system started to indicate that there was a risk attached to the money-losing piles – and this began after around thirteen plays, well before the conscious awareness of the chance of losing out.

This inner feeling – something you might identify as intuition or gut feeling – starts to influence your behaviour significantly earlier than when you make a conscious assessment. The same experiment was performed using participants with brain injuries that prevented their brains accessing the appropriate warning signs and the result was that they played the game far less effectively, even when they did become consciously aware of the downside of some piles. The intuitive, non-conscious influence was stronger than the conscious, logical assessment.

The same lack of pure conscious control can also be applied to our basic movements and response to senses. One significant indicator of this is in the work done by American neuroscientist Benjamin Libet. His experiments in the 1970s suggested that we do not become conscious of a stimulation until around half a second after the actual physiological experience – of, say, being touched on the arm – begins. Even more startlingly, he undertook experiments in the 1980s that seemed to move responsibility for at least some of our actions from the conscious into the subconscious mind.

The basis for this hypothesis was that when, for example, you decide to move your wrist, the activity in your brain apparently responsible for the movement begins a third of a second or so before you take the conscious decision to make that movement. The motion itself follows about two-tenths of a second later. It's as if the circuitry in your brain triggers the movement before the awareness of it filters through to your consciousness.* There is one major proviso here – the scientist's mantra, 'correlation is not causality'.

We will return to that in a moment, but before that I'd like you to take part in an experiment. It's not essential, but it really is worth doing. (If you've read my book *The Universe Inside You*, you may already have done this, but do try it otherwise.) To do the experiment you will need to go online. If it's not convenient now, you can do it later, but if you read on more than two pages, the experiment will no longer be valid. Go to the web page www.universeinsideyou.com/experiment9.html and follow the instructions. Afterwards, continue reading, as we'll come back to your experience.

* This anticipation effect was first observed by German scientists Hans Kornhuber and Lüder Deecke in 1964 and given the catchy name *Bereitschaftspotential*.

CORRELATION IS NOT CAUSALITY

When two events occur close together in time or in space, or if two values we measure move up or down together in time with each other, it's easy to assume that one event or value causes the other. However, it's equally possible that the direction of causality is the other way round, or that both have a common, different cause, or that the apparent connection is pure coincidence. Correlation is when the things happen together, causality is (as the name suggests) when one causes the other. A classic example of correlation without causality is the magical power of the accident hotspot sign.

It's not uncommon in the UK for road signs to be erected saying that there have been a certain number of accidents on the stretch of road ahead. After the sign is in place, the result is almost always a reduction in the number of accidents that occur. The assumption, by those who spend taxpayers' money on such signs, is that the sign has caused the reduction – money well spent. But in reality, exactly the same thing would happen if no sign were put up. This is due to a statistical effect called regression to the mean. Such signs are erected when there has been an unusually high number of accidents at a location. Statistically, such a run is most likely to be followed by a drop in numbers of accidents, bringing the overall average back to expected levels. There is no causal link between the sign being put up and the reduction in accidents, just a correlation – the two things happen to coincide.

This is a danger we face when companies and government bodies rely too much on big data – collecting vast amounts of data and getting a computer to look for correlations to predict the future. With enough data, it is always possible to discover correlations even if there is no causality whatsoever. If we take action based on processing vast amounts of data and searching for correlations, the result will be wholesale misinterpretation of that data. To show how easy it is, there's even a website that specialises in publishing such false correlations with the most unlikely apparent causalities.* For example, the divorce rate in the US state of Maine correlates very strongly with the US consumption of margarine, while there's an extremely strong correlation between suicides by hanging, strangulation and suffocation, and the US spending on science, space and technology. But few would blame the margarine or the spending.

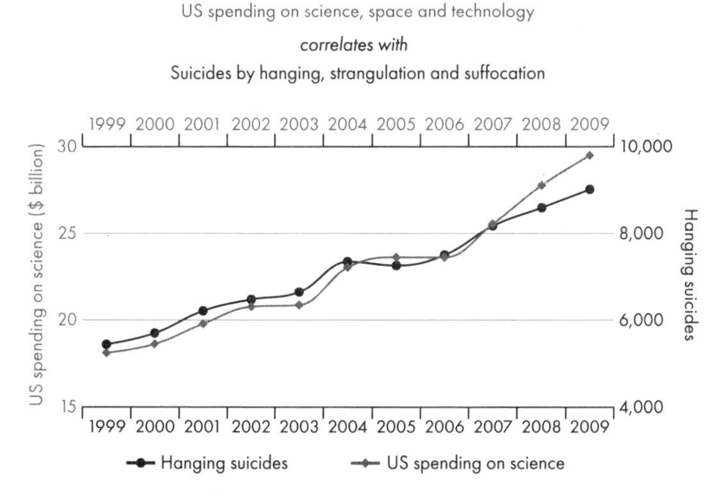

The remarkable correlation of science spending and suicides.
Source: Tyler Vigen

Similarly, it's entirely possible that the apparent correlation flowing from the subconscious intention to move to the moment of decision in the consciousness is not one of causality. This was suggested in a 2012 paper by American neuroscientist Aaron Schurger. He pointed out if there was an intention to move at some point – the participants, after all, knew what they were supposed to do – then it was entirely possible that the exact timing was triggered by random fluctuation in brain activity. When these fluctuations happened to hit a certain level, the individual made the

* www.tylervigen.com/spurious-correlations

conscious decision: that decision wasn't *caused* by the fluctuation, rather the conscious mind was merely given a poke to get working.

Another problem here is the subjective nature of the data collection in Libet's experiments. While it was possible to exactly time the motion itself, and the spike of activity in the brain, it wasn't possible to detect the conscious decision to act. Clearly, it was far too simplistic to ask the experimental volunteers (I assume they were volunteers) to shout out at the moment they make the decision to move. The very act of speaking itself is a motor action with its own inbuilt delays.

To get around this, Libet asked the volunteers to watch a clock and note which of a fast-moving series of dots was lit at the moment the decision was made. Even this, though, is fraught with difficulties. Think for a moment of a crocodile. When did you start to think about the crocodile? Could you stop thinking about that crocodile for a moment? It's hard to imagine that the volunteers could not think about what they were being asked to do. Assuming we accept the existence of consciousness, it is still profoundly difficult to pin down the point in time when a conscious trigger is made. Our perception of time is profoundly non-objective.

If, though, despite all this, we take Libet's findings at face value – that there is some kind of apparently subconscious brain activity, followed around 0.3 seconds later by the conscious decision, followed 0.2 seconds later by the action – what does this say for the concept of free will? Do we actually make conscious choices, or is our apparent consciousness merely the result of reacting to subconscious triggers? These are questions rather than answers because no one really knows.

As long as the concept of free will has been around (that's at least a couple of thousand years) there has been debate over its reality. This was intensified with the development of deterministic ideas based on Newtonian thinking,* which suggested that, as the French philosopher

* Determinism says that everything that happens at a particular moment in the universe is determined by what was happening the moment before – that the state that the universe (or some isolated part of it) is in, determines what will happen next,

Pierre-Simon Laplace noted, if you could have exact data on every physical object in the universe you should be able to perfectly predict the future – forever. This has proved a false concept, thanks to quantum physics, where uncertainty and the probabilistic nature of quantum processes means that the future can never be accurately determined.

Of itself, the reality of the quantum world does not provide a vehicle for free will – the future of quantum particles can't be predicted like clockwork, but they aren't subject to control either. However, we also know that most natural systems have sufficient numbers of complex interacting components that even if they were strictly deterministic, they would be impossible to predict. Described mathematically as chaotic, such systems include the weather, but also apply to biological systems like you to the extent that we might see a loophole opening for free will.

Libet came up with a specific way to give consciousness back a degree of responsibility for our actions. While he believed that the initial brain activity came before the conscious decision, he also claimed to show that the experimental subject could consciously veto the decision to undertake the movement before it happened – which feels related to Schurger's observation. Libet put forward a model where the brain initiates processes without conscious input, but it could then be overridden by the consciousness to moderate the resultant outcome.

WHAT LIES BENEATH

What is certainly the case is that there are both conscious and subconscious processes happening in the brain and it's perfectly possible for a sensory input, a memory and so on to be partly processed subconsciously before it enters our consciousness. You will almost certainly have had situations where, for example, you become aware of a sequence

....

leaving no choice available. *Que sera, sera*, and all that. Doris Day was presumably a determinist.

of sounds part way through. It is only at this point that the noises enter your consciousness. However, the subconscious processes were already aware of them, and you are then able to recall something that had not previously entered your consciousness. For example, if you become aware of a clock striking part way through its chimes, you can usually accurately say how many bongs have occurred.

If you didn't do the experiment that I asked you to do earlier, now's the last chance to head over to www.universeinsideyou.com/experiment9.html and give it a go. It's important you don't read on before doing so if you want to take part in the experiment.

In the experiment, a group of students are seen passing a ball back and forth between them. You were asked to count the number of passes made by the players wearing white. The correct tally is sixteen. But as you will have seen if you watched the whole video, we really don't care. The important point is that something surprising happened during the video. And around half of the viewers who are not familiar with this experiment will not have noticed it. Part way through, a person in a gorilla suit crossed the stage. If you didn't see it you are by no means unusual – if you don't believe it was there, go back and check it out.

The point of the experiment is that your conscious awareness of what's happening – your attention – is easily misled, presumably because you don't really 'see' the world outside projected on a screen in your head, but, as we have seen, your sensory inputs are determined by a series of brain modules that may or may not feed information into your consciousness.

As the later part of the video notes, the invisible gorilla is now a well-reported phenomenon and you may have been aware that there would be a gorilla, which makes it pretty well impossible to overlook. But it is still easy to miss the other two changes that occur. The most recent time I watched it, I was perfectly aware that there was something else in the video as well as the gorilla, and I still missed one of the two other events until significantly after it had already happened.

Whatever level of reality we ascribe to consciousness, there can be little doubt that it represents a relatively small part of what is going on at any one time in the human brain. As David Eagleman puts it:

> Although we are dependent on the functioning of the brain for our inner lives, it runs its own show. Most of its operations are above the security clearance of the conscious mind ... Your consciousness is like a tiny stowaway on a transatlantic steamship, taking credit for the journey without acknowledging the massive engineering underfoot.

Some philosophers go so far as to suggest that the limitations of consciousness mean that it is an illusion; others suggest that it does exist, but has no real function – your body and brain would work as well without it. According to them, consciousness gives a satisfying but unnecessary overview. And, since we're talking about philosophers, there are others still who feel that there is a phenomenon called consciousness that is more than simply the collection of your thoughts, senses, and experiences. In the end, it's not for me to tell you what you should think; it is for you to decide.

DEALING WITH THE UNREAL

If consciousness does exist, then this inner 'you' is able to access not only the inputs of your senses, thought processes and memories, but can also envisage what does not exist. You can deploy your imagination. I write fiction as well as science books.* When I am writing a crime novel, the characters and what they are doing are as real in my mind as are my memories of something that 'really' happened a while ago. Arguably, they are more real, because I can produce far greater detail about the

* See www.brianclegg.net/fiction.html

fictional situation than is the case with most memories, unless they are supported by photographs or notes. The same applies when I read fiction – I'm there, in the locale of the book, even though 'there' doesn't exist.

A very practical aspect of the ability of the inner 'you' to deal with fiction – and probably the reason that we are capable of it – is to be able to play 'What if?' This is at the heart of creativity and of what makes humans different from every other organism on Earth. Our extreme 'What if?' ability is why some still insist that we are the only animal that has the same degree of consciousness. There is no question in scientists' minds that this ability is anything other than the result of electrochemical processes in the brain. But it still feels like something that makes consciousness more than the combination of our memories, sensory inputs and thoughts.

This doesn't take us back to having an imaginary tiny 'you' sitting behind your eyes, experiencing the stream of consciousness as it flows by – but it would make consciousness something that emerges from the various component functions of the brain. As we have seen, such emergent phenomena do not exist independently, but can be far more than the sum of their component parts. Just think for a moment of what you are on a micro-scale. As simply a collection of atoms, or structures of molecules, or a large number of biological cells there is no you. That emerges from the way these components work together. Similarly, it could be argued that none of the component functions of the brain houses consciousness, yet it can still emerge from their collective interactions.

It ought to be stressed that even in the case of creativity, we can't ascribe everything to the workings of consciousness. When I ran business creativity seminars, I used to ask people in what circumstances they had their best ideas. It was never while sitting at their desk, trying (consciously) to come up with an idea, but rather when they were undertaking an activity that distracted them from conscious thought – anything from going for a walk to showering – or when they put the requirement to one side and left it alone by sleeping on it.

Your brain seems particularly good at making new connections and coming up with new ideas when not consciously focused on a requirement. But, equally, we can't dismiss consciousness from the process entirely. It is consciousness that asks the 'What if?' and 'How could I?' questions in the first place and makes something practical of the idea. Even so, there is certainly an unconscious component to creativity too.

THE UNBEARABLE LIGHTNESS OF LOGIC

There's another way that your problem-solving brain is influenced by what is built in below the conscious level. Try this simple problem. You are presented with these cards, each of which has a letter on one side and a number on the other. Which cards would you need to turn over to test out the hypothesis 'If there's an A on one side of a card there's an even number on the other side'?

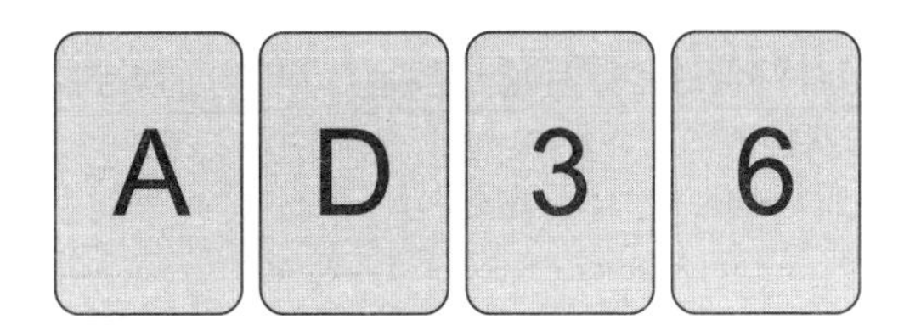

Take a moment to make your mind up.

The correct answer is that you need to turn over A and 3. You may have been tempted to turn over 6 – but the hypothesis would still be true whatever was on the other side. The requirement wasn't driven from even numbers, it was driven from there being an A. This makes it obvious that you need to turn over the A. Slightly less obviously, you also need to turn over the 3 – as should you find an A on the other side of it, the hypothesis would be untrue.*

* This problem is often stated without the proviso that the cards have letters on one side and numbers on the other. If that requirement is omitted we also need to turn over the D as there may be an A on the other side.

In part, the fact that over 75 per cent of people get this challenge wrong indicates that we're mostly not taught logic. But when the same test is set with cards showing alcoholic/non-alcoholic drinks in place of letters and drinkers' ages in place of numbers – with the proviso that there is an age on one side of each card, and the drink being consumed by a person of that age on the other – players are much better at knowing which cards to turn over to spot drinkers of alcohol under the legal drinking age.

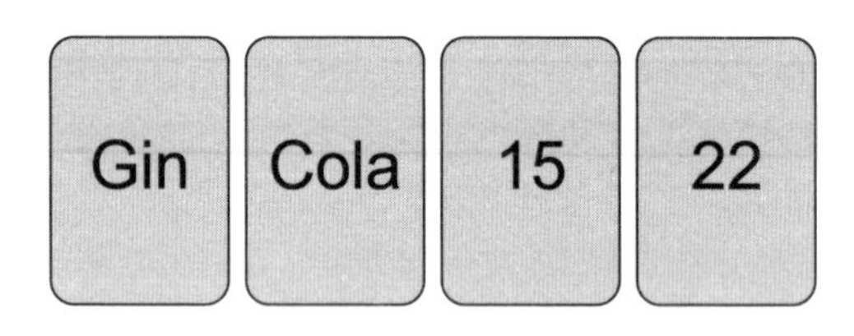

It's exactly the same problem, but your conscious attempt to solve it is aided by the subconscious linkage that makes it very clear there's no need to turn over the Cola and 22 cards. It has been suggested that this extra ability when the problem is moved away from the abstract is because mechanisms for dealing with social problems have evolved to operate without the need for the kind of difficult conscious logic necessary to solve the letter/number version of the problem.

A lot of the examples above have chipped away at the need for a conscious aspect of the brain at all – and, as we have seen, some psychologists and philosophers tell us that consciousness is totally illusory. David Eagleman suggests that this is a mistake, but an easy one to make because consciousness is not in charge, but is instead a fixing mechanism – something that takes over when the automated processes that deal with the vast majority of our actions fail to cope with a situation. It's the International Rescue of the brain, called out when disaster strikes.

This certainly makes sense in some cases, but it's a little harder to see why consciousness has to intercede when, for example, I decided a moment ago to pick up a tissue and blow my nose.* As shown by the Libet experiment, it's possible that the action was not triggered by my conscious mind at all. But if we accept Libet's explanation, presumably my conscious involvement was a check to ensure this really was the best action to take, adding the role of censor to rescue service.

YES, BUT WHAT *IS* CONSCIOUSNESS?

The very fact that there is nothing close to the kind of consensus found in physics or climate change in consciousness science tells us that, as yet, there is very little in the way of scientific evidence to pin down what consciousness is. We know *what* is happening, but it seems near-impossible to come to an agreement on *why*. English physicist Roger Penrose, in collaboration with American anaesthesiologist Stuart Hameroff, has proposed one potential physical source for consciousness, but to call it controversial would be a serious understatement. Before we explore the hypothesis, it's worth putting Roger Penrose into context.

Penrose is a classic eccentric genius of mathematical physics, who has made significant real advances in his field and is widely respected for them, while at the same time being considered something of a loose cannon who will support theories that don't necessarily have a lot of evidence to back them up. Although primarily a mathematician who works in physics, Penrose has always had an interest in perception and reality. It was Penrose who in his twenties, in collaboration with his father, the psychiatrist Lionel Penrose, came up with two of the best-known visual deceptions of the senses: the Penrose triangle and the Penrose staircase.

* Another interesting interaction of the conscious and subconscious brains is reflected in the way that some readers would have read this phrase as 'I decided a moment ago to pick my nose.'

Penrose triangle and Penrose staircase.

The staircase was the inspiration for M.C. Escher's famous image *Ascending and Descending*, where an example of such an impossible stair was incorporated into an elegant building. The linkage between Escher and Penrose would continue, as Escher's work exhibited a fascination with symmetry and the use of complex tiled surfaces, for example in Escher's *Study of Regular Division of the Plane with Reptiles*.

Roger Penrose would later come up with the answer to what might seem an almost impossible challenge making use of tiles: how to cover a surface with a simple set of tiles in such a way that the pattern never repeats. The patterns produced in Penrose tiling can, at first glance, seem to be structured in a regular, repeating fashion and can use as little as two apparently simple tile designs, but in fact never do repeat.

Penrose P2 'kite and dart' tiling.

Penrose has developed some impressive cosmological and astrophysical theory, though it is not necessarily capable of being observationally tested. But his collaboration with Hameroff raises the hackles of those who study consciousness. Penrose and Hameroff's hypothesis is based on parts of microscopic brain structures known as microtubules. These occur within neurons: Penrose and Hameroff, in what is rather grandly described as 'orchestrated objective reduction', propose that the microtubules host a form of qubit – the quantum computing equivalent of a computer's bit, and the interaction between these qubits is responsible for consciousness and free will.

A quantum computer is a real thing – it's a computer which, instead of having electrical bits that can have the values of 0 or 1, makes use of qubits – quantum bits – usually in the form of quantum particles such as electrons or photons. Qubits can be in more than one state at a time – for example, they can have a property that is, say, 40 per cent up and 60 per cent down. This enables a quantum computer with a relatively small number of qubits to handle problems of a complexity that would take a conventional computer the lifetime of the universe to compute.

Unfortunately, there is a serious problem with the practicality of Penrose and Hameroff's miniature quantum-computers-in-the-brain. The reason we can't buy a quantum computer off the shelf is that it is incredibly difficult to keep qubits in a functioning state and to get information in and out of the computer.* Most quantum computers have to be chilled within a fraction of a degree of absolute zero (–273.15°C or –459.67°F) or the thermal activity of the atoms destroys the qubits' fragile state. The idea of qubits functioning happily in the warm, wet environment of the brain is considered by most scientists to be extremely unlikely. There certainly is no evidence as yet to back up the idea.

You may also have come across New Age ideas that posit consciousness to be responsible for producing everything else – suggesting that there is no reality without the conscious mind. Developers of such concepts tend to make improper use of quantum physics. For example, there was an idea in the early development of quantum physics called the observer effect, which suggests that there needs to be a conscious observer to make real one or more of the possible outcomes of a quantum process. But this is an idea that has been long dismissed in physics and there is absolutely no evidence for it. This isn't science, it's science fiction.

DO YOU WANT TO BE BEAMED UP?

If Penrose's microtubule qubits exist, they would have to communicate with each other by a process involving quantum teleportation, which

* Technically speaking, if you have around $15 million to spare, you can buy yourself a D-Wave 'quantum computer'. However, this makes use of something that approximates to quantum computing, but technically isn't quite the real thing. At the time of writing, the best lab-based real quantum computers are briefly getting around 80 qubits operational – but are nothing like ready for general use.

definitely is a real thing in quantum physics, like a miniature version of a *Star Trek* transporter. The science fiction device disassembles the atoms of a person (or object), scanning them in such a way that they can be perfectly recreated after being beamed to a different location. To be able to do this, it hits up against a real limitation that we face when dealing with quantum particles such as atoms – making a measurement changes them.

This inability to look without changing things results in something called the 'no cloning theorem', which proves that it is impossible to measure exactly the parameters that define a specific quantum particle to be able to reproduce it elsewhere. However, quantum teleportation makes use of a clever work-around. It transfers one or more properties from one particle to another one, potentially at a distance. The second particle becomes identical to the first – but we never find out what the particle was actually like. The quantum teleportation mechanism depends on this.

At the moment, this teleportation is only typically done for a single property of a particle, rather than all its properties, and is performed a particle at a time or on a cloud of identical particles. Although it's possible it could be scaled up to deal with an object, or even a whole person, the challenge in doing so would be immense because there are a vast number of atoms in anything large enough for us to see.

The most commonly given figure for the number of atoms in an adult human is 7×10^{27}. Let's assume we put together a device that could process a trillion atoms a second. It would still take 7×10^{15} seconds to process a whole body. That's over 200 million years. So, while quantum teleportation is real, a *Star Trek*-style transporter is unlikely ever to exist. However, if one did, passing through it would involve being ripped apart with your individual atoms losing their current properties – total disintegration – while an identical equivalent was constructed elsewhere.

That identical copy would have all your memories, all your experiences, the exact same chemical and electrical processes in its brain. The copy you would be indistinguishable from the original you. Not just able to fool people – it would *be* you for an external observer. This has an interesting implication for the nature of consciousness. If consciousness were a total illusion, perhaps this wouldn't matter and you would be happy to use such a device. But if there is something there – a self with some kind of continuing existence (whether or not you believe in a dualist spirit), then such a transporter would wipe you out and produce a replica. I certainly would not be volunteering to try it.

Like the establishment of qubits themselves, quantum teleportation is not trivial to do in the kind of warm, wet surroundings we experience in the brain. It has been done outside the lab, but the devices dealing with it still had to be in an extremely controlled environment. Again, we've got a real thing but one that seems highly unlikely to be a mechanism involved in consciousness.

ARTIFICIAL INTELLIGENCE TO ARTIFICIAL CONSCIOUSNESS

Another futuristic consideration that science fiction played with long before there was any possibility of it happening in the real world was the idea of artificial consciousness – the thought that some kind of information technology would one day develop consciousness (or at least believe that it had done so, for those who think consciousness doesn't really exist).

Often, such stories depend on more and more capabilities being added to a network until it becomes so complex that it exhibits consciousness as an emergent property. This is the case, for example, in Robert Heinlein's classic 1966 novel *The Moon is a Harsh Mistress*, where the computer controlling the assets of the lunar colony becomes conscious and enables the colonists to obtain their independence. A rather less positive outcome is posited for technology gaining consciousness when the Skynet network gets ideas above its station in the *Terminator* movies.

There is very little danger that your laptop will suddenly develop consciousness. A computer has a very different and far less flexible structure than does your brain – the parallels between the two are often over-emphasised. But taking the lead from those science fiction stories, is it possible that computer scientists could actively build a device that was intentionally designed to develop consciousness? One serious problem here is the difficulty of distinguishing emulation from reality.

Ever since Alan Turing discussed a game where a player attempted to distinguish between a computer responding to questions and a human,* there have been computer programs written that try to emulate a human conversation. You can experience a very crude one of these called Eliza, originally devised in the 1960s, at the *Universe Inside You* website,† and the page also points to more sophisticated modern 'chatbots'. No one thinks that these programs are conscious. Even if one were developed that was perfectly indistinguishable from a human, we would know that underneath it is cleverly designed to emulate human behaviour.

However, American roboticist Hod Lipson, based at Columbia University, has a goal of making a truly conscious device. Lipson believes that a key to being able to give a computer awareness is what he describes as self-simulation. This would require a robotic device to have a mental model of its body and how it behaves physically, separate from the actual facilities required to control the body's movement.

Lipson has developed, for example, a robot arm with a self-generated model of itself which it has learned from interaction with its environment and which enables it to achieve tasks that it hasn't been explicitly trained to do. Although Lipson is not suggesting that the arm

* Although usually represented this way, Turing's idea was originally a variant on an old parlour game where the player attempted to determine if a correspondent was male or female depending on their answers to questions.

† www.universeinsideyou.com/experiment10.html

is conscious in the same way as a person, he credits it with internally being able to ask questions such as 'Where is my hand going to move?'

Like the chatbots, though, it is hard to see how it would be possible to distinguish simulation from reality. Even though the device has used 'artificial intelligence' learning techniques, rather than being hard-coded to perform in a particular way, this could simply mean that it has been trained how to write simulations. Lipson believes, however, that by extending this 'self-modelling' ability to cognitive processes, a version of consciousness could emerge.

WHAT IS IN CHARGE?

It should be clear by now that consciousness is not going to easily submit to analysis, and the philosophers and psychologists working in the field are probably not wrong to use that label 'the hard problem'. There remain many, many theories about consciousness which hover on the edge of science because there is rarely a satisfactory way to disprove such theories. As we have seen, some philosophers dismiss consciousness entirely. English psychologist Susan Blackmore, who has written widely on the subject, says: 'Consciousness is an illusion: an enticing and compelling illusion that lures us into believing our minds are separate from our bodies.'

It's certainly tempting from a rationalist, anti-dualist viewpoint, to suggest that consciousness doesn't really exist. But potentially that gives us problems when we think about crime and punishment. If we have no rational overseer, controlling our actions, but merely follow an intensely complex combination of low-level programs without any conscious decision-making, can we really be said to be responsible for any of our actions, however much these run counter to the law or what is accepted as common decency?

It could be argued that if consciousness does not exist, then you can't be held to blame for any of your actions. This doesn't mean wrongdoing would be ignored. It would be perfectly legitimate, should you commit a criminal act, to lock you away to prevent you reoffending, or to give you tools and techniques to avoid being in a situation again where you acted in this way – for example, through education. Equally, it would be acceptable to consider deterrence as a possibility, in terms of modifying the inputs to your unconscious subsystems and conditioning your behaviour. However, the concept of punishment and retribution would become totally meaningless.

As we will discover in Chapter 9, genetics can have a significant impact on behaviour and personality, as can brain injuries and tumours. There is a clear indication here of at best a limited role for consciousness in determining how you behave. We all make choices. Just because, say, you might have a gene variant that is more common in those who undertake violent crime does not mean that you will hurt someone. However, we do have to take in the big picture of our genetic makeup, our environment and the contribution of our mental subsystems as well. If consciousness does exist (which I am inclined to think it does), we must always remember it has a very limited influence over the brain's total activity.

ONE EXPLANATION OF MANY

Those working in the field offer a wide range of theories that try to provide an explanation for consciousness, often developed more in the spirit of philosophy than science. It feels pointless to recount all these theories (you can find more about them by checking out the books in the 'Further Reading' section at the back). However, to give a feel for the kind of hypothesis that can be put together, we can briefly explore an idea developed by the American duo of psychologist Jonathan Schooler and lawyer/philosopher Tam Hunt, based at the University of California, Santa Barbara. They put consciousness down to vibrations.

Anyone who has put a toe into what is known to scientists as 'woo' – New Age philosophies that use the language of science without any scientific content – will immediately have their woo detectors activated by that word 'vibrations', which is a favourite in the field when producing totally fictional theories (particularly when the theories are associated with crystals). But, to be fair, this isn't the case for Schooler and Hunt's idea. The starting point for them is that everything in nature – even the atoms in apparently still and solid objects – vibrates. This is definitely the case. These vibrations can link up, synchronising and supporting each other in a process known as resonance.

Resonance is why a wine glass will make a note when you rub a wetted finger round its rim. It's why the footfalls of pedestrians set the Millennium Bridge in London bouncing unnervingly until special dampers had been built into it. The tendency of things to pull together in particular frequencies of vibrations, synchronising, is called spontaneous self-organisation. Think, for example, of the apparent coincidence that the Moon rotates at just the right speed compared to its orbital velocity to keep the same face pointing towards Earth.*

The Moon's rotation has become 'tidally locked', because tidal forces distort the shape of the Moon, resulting in a greater gravitational pull on the closer part of the Moon's surface, which over time has synchronised its rotation speed with its rate of orbiting. For the theory of consciousness put forward by Schooler and Hunt, the synchronisation in question involves the different frequencies of activity of the various

---

* As opposed to the genuine coincidence that the Moon and the Sun have almost exactly the same apparent size in the sky, making total solar eclipses so dramatic. It's just that the Moon is about 400 times smaller than the Sun and about 400 times closer. But the Moon's orbital distance is (very) gradually increasing – eventually it won't be big enough to block out the Sun's disc entirely. One Oxford scientist has suggested that this isn't a coincidence, but an anthropic effect. He suggests that the tides arising from this Sun/Moon combination are beneficial for complex life forming, so we would expect to see it.

electrical oscillations in the brain, sometimes described as 'brain waves',* which, depending on the type of activity of the brain, can more or less synchronise.

Schooler and Hunt suggest that what we experience as consciousness is the resonant synchronisation of the vibrations in the components of the brain, producing a whole that is more than the sum of its parts. The oddest conclusion they draw from this theory is that absolutely everything has a degree of what we call consciousness – even, for example, a grain of sand. But the process is acting at such a low level that it has no measurable effect, whereas the unparalleled complexity of the human brain takes the concept of consciousness to a new level.

Is there any evidence to support this theory? Not really. Remember 'correlation is not causality'. Just because there may be more resonant vibrations going on in your brain than in that sand grain doesn't mean that this is the cause of you being more conscious than a rock – or, for that matter, that consciousness even exists at all. Unlike most pseudo-scientific thinking, though, Schooler and Hunt are proposing a scientific mechanism – there simply isn't evidence to either support or counter the theory. And that's what the many respectable† theories of consciousness generally feel like.

I argued earlier that at least one role of consciousness is to contribute to creativity. Now we need to turn this around. This book is about what makes you *you*. Inevitably a lot of this arises from your biology and how that biology came about – but by no means all of it. Human creativity means that we are able to shape our environment and change our abilities in ways that are unavailable to any other organisms. It's time to go beyond biology.

* Nothing to do with having a brainwave, these are oscillations detectable in the electrical activity of the brain.

† Perhaps 'respectable-ish' would be more accurate.

8

LIFE IS MORE THAN BIOLOGY

For the vast majority of our pre-human ancestors (as is the case for most other living organisms today) life was entirely about the basics of biology – keeping alive and reproducing. However, over 3 million years ago, the predecessor of *Homo sapiens* started to use stone tools, adding an extra dimension to the biological definition of what they were. Now our lives have become a complex web of technology and trade, without which few of us could live for more than a few days.

The media is constantly worrying about the impact of technology on what we are. Is social media making us less able to socialise in the 'natural' way? Does playing violent video games make us more aggressive in the real world?* These are certainly questions that need to be asked, but they are only a very small part of the impact that human creativity has had on what makes you the person you are.

Biologically speaking we are very similar to the first humans around 200,000 years ago, which means that sometimes our reactions to the world are out of sync with its current nature. But it would be naïve to suggest that you are no different from your distant ancestor. We have

* The idea that violent video games make people more violent is a constant in the media, but there is no good scientific evidence to suggest that this is the case. At the time of writing, the idea has just been dragged up again by US President Trump in response to the latest mass shootings in his country.

undergone small evolutionary changes, it's true – but the biology side of the equation is minimal. It is that new technological world that we have created, for good or ill, that makes you very different from your predecessors. We are now able to transform our environment and our capabilities in ways no organism before us has ever been able to do. And perhaps the most dramatic way that this ability has influenced what you are is that without it, you would have a more than 50 per cent chance of being dead.

THE BENEFITS OF NOT BEING DEAD

Enthusiasts for returning to a simpler, bucolic, medieval way of life tend to gloss over the reality that for most people prior to the industrial revolution, life was unpleasant. Work was physically extreme and rarely intellectually stimulating. Without the ability to transport food from afar, a ruined crop was not just a financial worry – it meant the difference between life and death.

Even now, with all our technological benefits, agriculture is at the mercy of natural forces. I was recently having a drink in the pub with a couple of farmers, one of whom described how his entire crop of turnips – several acres – had been wiped out by swarms of a tiny insect called a flea beetle. He knew others who had lost hundreds of acres of crops to these pests. This didn't exactly make him happy, but his family was not going to go hungry as a result of it. However, things were very different in the past, when losing a crop meant starvation.

Food was not the only problem before technology transformed our environment – it is only in the last 150 years that medicine has gradually become a scientific discipline. Even in Victorian times, it was not unusual for doctors to do more harm than good by imposing treatments such as bleeding that harmed rather than helped the patient, as they were totally ignorant of the causes of disease. The combination of food shortages, cold and damp homes and the inability to counter disease, resulted in the depressing statistic that at the start of the 20th century in the UK, average life expectancy was around 45 for men and 49 for women. For a century or two earlier, you can reduce those figures by a good ten years.

We have to be a little careful here, because it's easy to get misled by statistics. Averages are very useful, but they don't tell us enough on their own to understand what is going on. Imagine I took you to a full-to-capacity Wembley stadium in London and told you that the average wealth of the people in that stadium was £1.2 million ($1.5 million) per person. It would be easy to assume the place was packed with millionaires. What, though, if one of the people happened to be Jeff Bezos? At the time of writing the richest person in the world, Bezos is worth around $154 billion. That would mean that the other 89,999 people in that packed stadium were totally penniless. When extremes are involved, the average gives a misleading picture.

Similarly, those short life expectancies from earlier times don't make it clear what was happening. It's not that the majority of people died in their mid-thirties or forties. Instead, if you made it to 40, you may well also make it to be 60. But far more people died young than do now. Exact figures are hard to come by but go back to 1800 and more than 40 per cent of children died before they were five. Less than half survived to adulthood. Think about that for a moment – back then, the *majority* of funerals would have been for children.

So, assuming you are already at least eighteen, it is more likely than not that in a past time, without our technological safety net, you would have been dead by now. It's hard not to accept that being alive rather than dead is a significant component of what makes you what you are today.

AROUND THE CAMPFIRE

Perhaps the first important contributions to reducing the chances of dying, predating medical treatment, were the use of basic tools and fire.

The benefit of the tools, such as stone axes, are fairly obvious – they made killing animals for food and defence, collecting food and later agriculture more practical. However, fire would also have played a major role. It's not just a matter of keeping people warm and scaring off predators – fire made it possible for our ancestors to cook food, which would prove to have huge benefits.

It's probable that the initial benefits of heating up food, no doubt discovered as a result of an accident, were those of improving flavour and texture. Cooking modifies some of the proteins in foodstuffs, making them far easier to chew and to digest. Those who take on a diet that consists only of raw plants soon discover that they have to spend a huge amount of their time simply chewing on the food to get enough nutrition. Similarly, cooking transforms flavours and odours, breaking down carbohydrates to simpler sugars and releasing pleasant-smelling chemicals that enhance the experience of eating the meal.

Beyond the edibility benefits, though, came the health benefits of cooking. In recent years 'raw' has come to be a positive term, to contrast with the negatives associated with processing food. However, cooking also kills off bacteria and other dangerous contaminants, and destroys a range of natural poisons. We need to remember that many of the most poisonous substances in existence are natural. So, for example, kidney beans contain phytohaemagglutinin, a deadly substance that is, thankfully, destroyed by cooking.

THE FOUR HUMOURS

Effective medical ideas took far longer to develop than cookery skills. Like astronomy, medicine was long locked into an inaccurate model of reality. Just as astronomers struggled to move away from an Earth-centred universe, so medics in many parts of the world settled on the idea of having four 'humours': substances within the body which had to be kept in balance, often along with some idea of a life force that was channelled through different parts of the body. The humours – black bile, water, blood and yellow bile – corresponded to the four elements that were believed to constitute matter at the time: earth, water, air and fire respectively. Sadly, though, the theory was totally fictional.

This meant that early medicine had very limited benefits. Usually there were some herbal remedies that did happen to work, discovered by trial and error – but medical theory was based on a totally spurious concept, generating plenty of treatments that were useless or, more likely, harmful. This didn't mean, though, that there weren't some sensible ideas on basic good health.

In a fascinating document known as *Letter Concerning the Marvellous Power of Art and Nature and Concerning the Nullity of Magic*, the 13th-century English friar Roger Bacon spent some pages looking at the care of health. Admittedly Bacon was distracted by the myths of the time. He noted: 'The Lady Tormery in England, while searching for a white hind, found an ointment with which the keeper of the woods anointed his whole body except the soles of his feet – and he lived three hundred years without any corruption save pains and suffering in the feet.' However, Bacon's broad advice on health would not be amiss on an NHS website. He advises that a 'real remedy' might be found if 'a man from his youth would exercise a complete regulation of his health in all manners pertaining to food and drink, sleep and waking, movement and rest ...'

By the 17th century, medical support was provided by a combination of apothecaries dealing in herbal medicine, physicians whose work was based on the four humours theory, and surgeons, who started pretty much as barbers on their day off – it's not entirely surprising that things weren't great. It was only with the move away from the idea of diseases being caused by miasmas (foul air, as opposed to the 'fresh air' people were encouraged to get for their health) that medicine could start to make steps forward.

Though still not perfect, medical science has made huge strides through better understanding of how the body works at the molecular level, making it possible to develop far more sophisticated drugs, as well as understanding viruses and bacteria and having surgical techniques that would have been impossible without modern technology, especially anaesthetics. It's very unlikely that you will have got to the point of reading this book without having had some medical intervention in your life that helped you survive to be what you are today.

YOU ARE WHAT YOU WEAR

What makes you *you* is not all about survival, though – it can also be about distinctiveness. When we first see someone, what they are wearing makes a significant impact. Our clothes say something about us. And though it's true that clothing can also be something that helps us stay alive – particularly when we venture into extreme environments (think of spacesuits, thermals or wetsuits, for example) – clothes are far more than this, contributing to what makes us individuals and members of tribes.

As noted in my earlier book *The Universe Inside You*, woven cloth dates back at least 27,000 years and needles for stitching cloth or animal skins around 40,000 years. We also found that it has been estimated that human beings have worn clothes for between 50,000 and 100,000 years, thanks to an investigation of the origins of the body louse.

Head lice have been around far longer than body lice, and it's thought that the body variant of the louse family was only able to stray away from the protection of the hair when we started to wear clothing – timing that fits well with the estimated timescale for humans moving out of Africa into colder environments. Although clearly wearing warm furs would have helped keep those early humans warm, there is strong evidence that other clothing was worn more for effect than for insulation.

A good example comes from a statue that is around 25,000 years old, found in a cave at Lespugue in south-west France. Carved out of a tusk, the small statue of female wears a sort of skirt, hanging from below the hips. From its position and construction from twisted fibres, the skirt seems far more likely to have some sort of ritual or symbolic purpose – or simply to look attractive and indicate personal status – than to have any role in keeping the wearer warm.

Anthropologists point to examples where clothing is definitely more symbolic than functional to emphasise how much what we wear establishes our status. In Tudor times, for example, it was illegal for those of low station to wear elaborate clothes. Even now uniforms, for example, play a status role, as do the impractical horsehair wigs of the British judiciary. Other items of clothing function as stand-ins for sexual displays (think of the Tudor enthusiasm for large codpieces) or as a mark of unavailability, as with the Islamic hijab and the medieval Christian wimple. Even standard items of dress from mass manufacturers can help bolster our identity. Personally, for example, I like to wear Dr Marten shoes. This isn't because they are the best shoes, or provide any survival benefit – it's simply part of my personal concept of identity.

Some of us establish our individuality in the way we dress – or, for that matter, in the way we modify our bodies. This might be through temporary changes, such as makeup and jewellery, through the semi-permanent means of piercings or through long-lasting tattoos. All such adornment tends to be subject to fashion. Tattooing has seen a recent change to become far more generally acceptable – though an earlier attitude to tattoos, which meant that you pretty well only saw them in the UK on sailors 50 years ago, was equally a relatively recent trend. The oldest known tattoos date back around 5,300 years, found on the person of Ötzi, the so-called iceman. Ötzi's body was found by a tourist in 1991, frozen in a glacier between Austria and Italy. Ötzi had a total of 47 tattoos on his body.

Although many of us still avoid tattoos because of their lack of flexibility – few tattoo wearers would wear the same clothes their whole life – makeup and body painting have been far more widely employed as far back as history allows us to explore. Sometimes this is a matter of covering up flaws, sometimes embellishment – but many wearers of makeup would consider it a significant part of what makes them who they are.

An example of the degree of effort going into this modification is the use of eye makeup by the ancient Egyptians. The oldest known cosmetic artefacts are Egyptian eye makeup palettes, dating back around 12,000 years. Commonly, this involved the use of kohl, a black substance based on soot mixed with minerals. It has been suggested that kohl was partly used to reduce the amount of bright sunlight reflected from the skin into the eye in very sunny conditions, though sadly any health benefit gained from this would be more than countered by the presence of those minerals, which often included lead sulphide, resulting in dangerous absorption of lead into the skin.

Similarly, human beings have changed their hair colour for thousands of years. The ancient Greeks, and following them the Romans, had a thing for blond hair, reflected in the comment of Pope Gregory the Great familiar to anyone who learned Latin at school: when he saw fair-haired slaves from the Anglian kingdom in Britain, he is said to have remarked '*Non Angli sed angeli*' – not Angles, but angels. Ancient Greeks, particularly, made use of a 'sun-in' hair bleach that involved washing their hair in a special potion and sitting in the sunlight while their hair grew lighter. In a more modern form, dying hair is a way that you may have also gone beyond your biology to make the 'you' that you see in the mirror.

IT FROM BIT

There's a rather obscure cosmological concept that was labelled by the American physicist John Wheeler 'it from bit'. It's the idea that the whole universe is, effectively, constructed from information. Whether or not that is true, it's hard not to see information technology as one of the main drivers that makes us what we are today. Of course, there are plenty of other contributory technologies and capabilities – transport, for example. However, it's hard to think of another development that has become so transformational so quickly, and continued to be enhanced for so long.

Your default image when confronted with the words 'information technology' is probably a computer or a smartphone. However, we need to bear in mind the huge importance of writing, the most fundamental of the information technologies, in the development of our societies, economics and science. It's writing that helped prevent us from reinventing the wheel, making it possible to pass on knowledge from place to place and time to time.

One of the most famous quotes from a scientist – Isaac Newton (in a soundbite that he probably paraphrased from the English scholar and author of *The Anatomy of Melancholy*, Robert Burton) – is 'If I have seen further it is by standing on the shoulders of Giants.'* But without books and letters, Newton would not have been able to build on the ideas of others.

Early information technology went through a number of modifications. Before electronics, the two biggest developments in book tech were the move to the codex and printing. Early books came in the form of scrolls. These were limiting in length, clumsy to store, hard to hold when reading, and impossible to easily shift position in, back and forwards. Although the Romans made practically no contribution to science, their technology was impressive and one of their particular contributions here was the codex, sheaves of sheets of paper bound together to form what we would now regard as the traditional paper book.

Printing with moveable type, too, was essential to allow for easy transmission of information. For hundreds of years, the only way to

* Newton's famous phrase, written in a letter to Robert Hooke, is often used as an inspirational quote. It's worth noting, though, that Newton probably intended this to be a veiled insult, as Hooke had claimed that Newton stole ideas from him and Hooke was anything but a giant.

reproduce a book was to laboriously copy it out by hand. The printing press changed all that. Printing had been around for some time, originally performed by carving a page out in reverse on a piece of wood and printing a whole page at a time. The earliest known existing book to be printed this way was the Chinese *Dunhuang Diamond Sutra*, dating back to 868.

The Chinese were also responsible for the first invention of moveable type, where the individual characters for the page are placed on separate small blocks which are bound together to make a pageful; after printing, they can then be rearranged and reused. The type blocks started out as wood or ceramic, dating back as early as the 11th century, though the more durable and more easily shaped metal was taking over by the 14th century.

Despite inventing the technology, it was not in China that moveable type really sparked a revolution, but in Europe. This was probably because of the much simpler character set needed for European printing. When printing using the Roman alphabet, fewer than 100 different type characters were needed – but for a book in Chinese, with potentially thousands of different characters, the benefits of moveable type were far smaller.

WORDS AT THE SPEED OF LIGHT

Although books and journals were absolutely essential to the spread of knowledge, we tend now to think of information technology as electronic, and the first electronic revolution would come not with the computer, but with a transformation of communication.

Throughout known history, information had mostly travelled at the speed of animals, whether carried by a walking human or a messenger on horseback. It could take days, weeks or even months for a message to get from one part of the world to another. Admittedly, there were ways to speed things up – but they were limited in the complexity of the information that could be transmitted. The two media that have always been used in a small way to do this are sound and light.

Even without technology we make use of both of these, communicating by speech and visual indications from gestures to body language. But with simple techniques it was possible to extend direct human communication. Sound was employed, for example, in using church bells to ring out a summons to church or to warn of invasion, while light enabled a simple alert to be transferred a few miles at a time using smoke, flags or night-time fires. Where necessary a chain of stations could be used, passing on the visual signal from place to place in line of sight.

While definitely a means of communication, such mechanisms were not going to change everyday life much. It took the ability to transmit a series of symbols, some sort of code, to make serious headway in transmitting information at speeds over and above that of the dispatch rider's pouch. Although it's likely that such ideas were used on a small scale earlier without making it into history, one of the earliest practical examples was developed by French inventor, Claude Chappe.

Chappe's first attempt was not likely to go down well with the neighbours. Sender and receiver both had a clock with a second hand. The sender also had a large gong.* A number of pre-arranged bangs of the gong were used to synchronise the clocks. Once this was achieved, a message could be sent by sounding the gong when the sender's second hand lined up with a particular number on the clock face. To make this even easier, sender and receiver could paste a series of letters onto their clocks.

Sound has the advantage over light of not having to travel in a straight line, but it can't get very far and is susceptible to variations in range depending on the wind. To transmit a message any distance this way would have meant having repeater gongs every half-kilometre or

* Chappe actually used a marmite – not the British malt-based spread, based on brewery waste products, but the large metal French cooking pot that the spread was named after (a marmite is shown on the Marmite label).

so – not to mention causing real irritation to local inhabitants. So, having got the basic concept together, Chappe switched to using the less intrusive medium of light.

This really increased the range that was possible between stations – to 10 or 12 kilometres – and it was pleasantly peaceful. Chappe's first design had a wooden panel that could be rotated to show a dark or light side, but he standardised on something closer to a giant human, waving semaphore flags. He did this by mounting two large wooden arms on a tower (to give better visibility). Each arm ended in a rotating section, so the combination of position of the arms and of the end pieces could spell out a range of letters. By adding lamps to the moving parts, the device could even pass on a message in the dark.

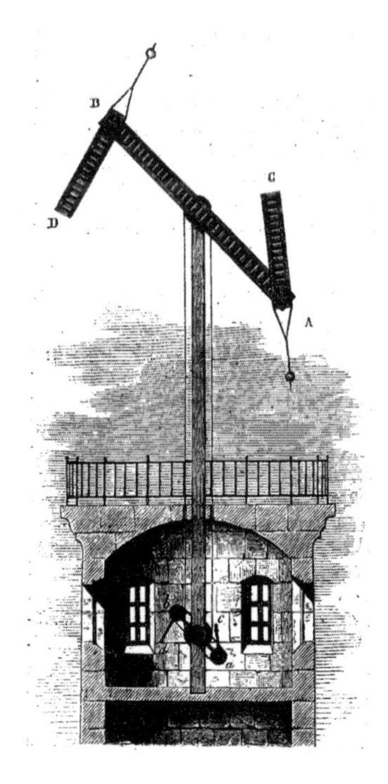

Chappe telegraph station.

Chappe wanted to call his invention the *tachygraphe*, roughly, 'fast writer' in Greek, but a friend considered this an uninspiring word and suggested instead *télégraphe*, or 'far writer', which became the familiar name for the technology. Within three years, by 1794, a string of Chappe telegraph stations had been built across France from Paris to Lille. The fifteen devices, covering between them a distance of 210 kilometres (130 miles) could send in minutes a message that previously would have taken a day to deliver on horseback.

Over the next 40 years, more than 1,000 of these telegraph stations (known as semaphore stations to English speakers) would be set up around the world, but by then the approach was already under threat from a less clumsy technology. It was time for electrons to take over.

DOWN THE WIRE

The decisive year for the electric telegraph was 1844. Experiments had been underway for some years, both in the US by sculptor Samuel Morse, who began work in 1832, and in the UK by William Cooke and Charles Wheatstone, who started a couple of years later. Morse would send the first message using his eponymous code on 24 May 1844, dot and dashing out the portentous phrase 'What hath God wrought', transmitting down a wire alongside the railroad track from Washington to Baltimore.

Cooke and Wheatstone took the less flexible, but easier to use route of putting an indicator board at each end of the wire. The positions of a series of pointers on these boards were synchronised by the electrical signals, so they could be used to spell out messages a letter at a time. Cooke and Wheatstone chose to lay their first line from London along the Great Western Railway to Slough, around 32 kilometres (20 miles) to the west.

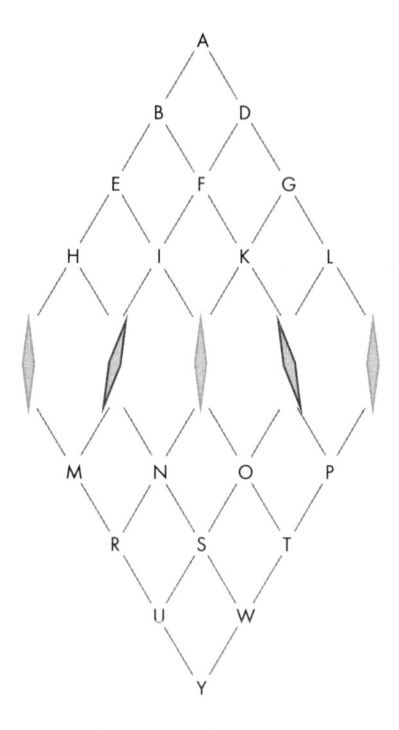

In this five-needle version of Cooke and Wheatstone's telegraph the angled needles are indicating letter F.

It was no coincidence that both telegraph pioneers transmitted alongside railway lines. In part this was for ease of laying the cables. Unlike the twisty roads of the period, railway lines were relatively straight and had dedicated land to either side, without the interference of crossroads and other problems that beset the highway. It was also the case that the railways had started the demand for a rapid form of communication. Before railway travel, each town had its own time, which could be minutes away from the time elsewhere. But a train timetable demanded uniform time – and the ability to synchronise clocks from place to place.

This doesn't explain Cooke and Wheatstone's specific choice of Slough as their location to connect to London; at the time, Slough was best-known as the home of the astronomer John Herschel. The house had been built by his father, William, who had moved there to be near to the king's residence at Windsor Castle. Slough, similarly, was a key location on the railway, providing as it did access for both Windsor Castle and the prestigious school at nearby Eton.*

The choice of Slough proved to be a very convenient one for publicity purposes, as, on 6 August 1844, Victoria's second son Alfred was born at Windsor. According to reports at the time, *The Times* newspaper made it onto the streets of London carrying the happy news within 40 minutes of the announcement being made in Windsor, thanks to the telegraph. Ten years later, the same link from Slough to London would result in the capture of a murderer.

On 3 January 1854, a man named John Tawell murdered his mistress in Slough and fled the scene, taking the train to London. Once he was on the train and away, he must have assumed that he was safe to disappear into the streets of the capital, as information about his crime would not have reached London yet. However, the Slough authorities managed to get a message sent down the telegraph line to Paddington station. Tawell was wearing a long brown coat, which enabled him to be recognised as he got off the train, the Slough office having sent the message to apprehend a man 'dressed like a kwaker' (the Cooke and Wheatstone system didn't have a letter Q). Tawell was tried and hanged for the crime.

In essence, this was an example of the information technology being used to bring time in the two locations into line with each other. The same effect would also be used to perform a crude piece of Victorian time travel. Bookmakers always like to take bets up to the last minute. Back then, when it would take hours for a result to arrive, say, in London from a race that took place 320 kilometres (200 miles) away in York,

* The Eton school's headmaster had demanded the station be built at Slough rather than Eton itself to make it harder for the boys to escape to experience debauchery in London, which is why Slough had a far grander station than was required for the small town it was at the time.

the bookmakers happily continued to take bets well after the race had been run.

Once the telegraph came into operation, clever betters would arrange for the result to be transmitted to them, so they could make a big win with no risk of losing money. The telegraph companies quickly became aware of this and refused to transmit betting information – messages had to be sent by company agents and could not be dispatched directly by the public, enabling such censorship – but would-be betters got around this by using coded messages. Eventually, the bookies dropped the practice of accepting bets after the race, closing the loophole.

INHUMAN COMPUTERS

As we have seen, there is talk these days of the possibility of computers developing a form of artificial intelligence; but originally computers had the conventional kind of intelligence – because a 'computer' was a person who carried out computations by hand. Such computing was tedious, but essential, for everything from astronomical work to creating tide tables. It was while helping out his friend, the astronomer John Herschel, in the computation of astronomical tables in 1821 that English inventor Charles Babbage is said to have cried out, 'I wish to God these calculations had been executed by steam.' Babbage was also encouraged by the work of a French mathematician, Gaspard de Prony, who, inspired by the writings of the economist Adam Smith, had experimented with mass production of tables of logarithms, breaking down the task as a machine might.

Whether inspired by his painful experience or by Prony, Babbage began to work on the idea of a mechanical computer. Of itself, the idea of a mechanical calculator was nothing new. Such devices go back at least as far as the Antikythera mechanism, recovered from a Greek shipwreck and dating back to the 1st or 2nd century BC. This device has a complex mechanism of gears, enabling it to act as a specialist analogue computer, predicting the motion of heavenly bodies and more. And there is a more direct antecedent of Babbage's work in calculating machines such as the one devised by French mathematician Blaise Pascal in the 1640s, a number of which were constructed.*

Like the Pascal machine, Babbage's first concept, the Difference Engine, was a mechanical calculator based on a series of gear trains, but far more sophisticated in the range and scale of its calculations than its predecessor. Babbage constructed a section of a Difference Engine, around one-seventh of the total machine, but never completed a working device.† His failure to finish the Engine was (not surprisingly) a great irritation to the British Government, which had put over £17,000 into the project.‡ Despite government complaints, however, Babbage dropped the Difference Engine for a far grander idea, the Analytical Engine, which was inspired by an ingenious French industrial development.

The inspiration here came from the sophistication of the silk weaving industry. Making any kind of complex pattern with this extremely fine thread was painfully slow – so much so that two loom operators might between them only produce an inch of material a day. This limitation had not escaped French engineers, and in the 1740s, a government factory inspector called Jacques de Vaucanson devised the silk loom equivalent of a musical box. Just as the pins on the cylinder of a musical box trigger notes as they pass by, Vaucanson's invention used pins to control the different-coloured threads. This was certainly an advance, but each cylinder was expensive and slow to produce – and the process was limited by the size of the cylinder. One turn, and the pattern began to repeat.

* Allegedly, Pascal's calculators were very unreliable.

† Two full examples were constructed in the 1980s, proving the effectiveness of Babbage's design.

‡ In current money this would be around £1.2 million as a cash amount or £13 million in terms of labour value.

Before the Vaucanson cylinder could become commonplace, it was displaced by the idea that inspired Babbage. A previously unknown figure (by all accounts, something of a vagabond), Joseph-Marie Jacquard, brought flexibility to the programming of the technology. He laid out the weaving pattern as a series of holes on cards, the holes indicating whether or not a particular colour should be used at that point. Because the train of punched cards could be as long as the pattern required, any piece of weaving could be automated this way. Before long, Jacquard looms were turning out not one inch of silk material a day, but two feet – a remarkable transformation of productivity.

The punched cards in a Jacquard loom.

However, the power of the system was not just in its speed – it was the flexibility of its information storage that appealed to Babbage. A treasure that he often exhibited to visitors was a portrait of Jacquard that appeared to be an etching – on close examination, the picture was woven from silk with a remarkable 24,000 rows of thread making up the image. Such a product would have been impossible without

Jacquard's technology, and Babbage realised that a similar approach could be used in a truly revolutionary computing device: his Analytical Engine.

Dismissing the fixed, geared relationships of his earlier design, Babbage wanted the Analytical Engine to have the same flexibility that was demonstrated by the Jacquard loom. For the Difference Engine, the data to be worked on was entered manually on dials, and the calculation was set by the architecture of gears. In the Analytical Engine, both data and calculation would be described by a series of Jacquard-style punched cards, allowing for far greater flexibility of computation.

There was just one problem. Although Babbage designed the Analytical Engine in concept, he never managed to construct even a part of it – indeed, it is unlikely that his design could ever have been successfully built with the technology of the day. Even so, his plans were sufficiently detailed that someone was able to speculate on how the Analytical Engine might have been used, creating our modern-day image of Ada Lovelace.

The then Ada Byron, daughter of the famous poet, had first met Babbage when she was seventeen and shared his many enthusiasms. There were even hints of a possible marriage, though it never had much of a chance, as Ada Byron's mother ensured that her daughter was married to a man with better prospects, William King, later the Earl of Lovelace. Ada Lovelace certainly wanted to work with Babbage, and made a contribution to the documentation of the Analytical Engine by translating into English a French paper on it by the Italian scientist Luigi Menabrea. Lovelace tripled its length with her notes, including some suggestions of how the Engine could be applied. She even included a few examples of potential programs for the (non-existent) computer. All but one of these had already appeared in Babbage's lectures, but one appears to be original.

Unfortunately, Babbage's engines were a technical dead end. The next development in computing would be an apparent step backwards to the idea of punched cards without the clever analytical engine to work on them. This technology was devised by American inventor Herman Hollerith, who used electromechanical devices to sort and collate punched cards, an approach that rescued the US census, which was in danger of taking longer than the ten years between censuses to process manually.

Hollerith's Tabulating Machine Company became International Business Machines, which was eventually reduced to its initials, IBM. The true electronic computer came out of the Second World War with the development of Colossus at Bletchley Park in the UK and ENIAC at the University of Pennsylvania in the US. Since then we have seen the accelerating route from valves (vacuum tubes) to transistors, from transistors to integrated circuits and from computers the size of a building to pocket computers in the form of a smartphone, plus the parallel development of communication technology in the internet.

TRANSFORMING EVERYDAY LIFE

To begin with, information technology was primarily used behind the scenes, with computing taking place in universities, government departments and big business. When the concept of the personal computer first emerged in the 1970s and 80s, a considerable amount of effort had to be put into trying to persuade potential customers of the point of having a computer in the home. Games were an early opportunity to sell the idea, but the manufacturers wanted to move beyond what was seen as a childish application, to make the computer part of everyday life. They resorted to feeble suggestions such as 'You can use a computer in your kitchen to store recipes.' Arguably the reason computers shape our lives outside work so much today was the addition of the internet.

The internet itself dates back to the 1970s, when it was developed in the US as a way to connect a terminal (effectively an electrified typewriter) to a remote mainframe computer so that the computer could be used from elsewhere. What started off primarily as a military and university network (the military separated off into their own network fairly early on) initially made relatively few inroads into either the business or the home market.

In the mid-1990s I was writing for computing magazines, and as a result attended the London launch of Microsoft's shiny new Windows 95 operating system. It was a glossy affair in a Leicester Square venue, but one thing is particularly interesting in hindsight. During the Q&A, I asked what support the operating system would have for the internet. The answer was that the internet was primarily of academic interest, not of any real concern to Windows customers.

At the time, Microsoft was determined to follow the likes of CompuServe, AOL and Apple in producing a proprietary network which they called MSN. Within a few years, though, driven by Tim Berners-Lee's addition of the World Wide Web on top of the basic internet architecture, things would have radically changed. Berners-Lee only intended the web to be a mechanism for academics to share information, but its combination of hypertext (an earlier concept from the 1960s that had never before been properly implemented) and the ability to see pictures and text from around the world changed everything.

Now, like millions of others, my business is largely conducted via the internet, as is so much of my home life, whether it's connecting with friends on social media, finding a country pub to visit or making use of GPS data to get me to my destination in the car. Watching the 1980s-set TV show *Stranger Things* (via the internet-based Netflix) a while ago, I was struck by the way that the characters had to look up information in books. Of course, we still use books – you are demonstrating that by reading this one (although you could be using an internet-enabled Kindle device to read it). However, the instant access to such vast amounts of information is something that has hugely changed how we live, and that is still to be truly absorbed into, for example, the way that we educate children.

Information technology is, as we have discovered, by no means the only transformative technology that has helped make humans what we are today, but it makes a great example of why we cannot only consider your biological makeup in deciding what makes you *you*.

THE HAPPY ACCIDENT

Arguably, creativity distinguishes the human species more than any other characteristic and enables us to modify our environment using technology (as well as making it possible to modify the mind with art and literature). Creativity is all about finding new connections, turning existing thinking on its head and exploring the untried. Even if you don't think of yourself as creative, it is a major part of your makeup. Often, innovation is purposeful, performed in a guided fashion. But some of the best, most important and simply most fun aspects of creativity in human history have been the result of accidental epiphanies.

Accidental epiphanies, when an idea emerges from an accidental discovery, a mistake or seeing something differently, have been important to our creativity as long as humans have noticed unexpected aspects of the world around them and responded to them. While it's true that much creativity and discovery is a result of Edison's quoted '99 per cent perspiration', or of building logically on existing knowledge, it's surprising how often the development of science and technology has been driven this way. And such happenchance findings brought us a range of early ideas and technologies, as we have seen with fire and its spin-offs such as cooking. However, the pace of serendipitous innovation would take off hugely from the 15th century with the invention of discovery.

When Columbus attempted to sail west to China and instead hit on the New World, few had an appropriate word to describe what he had done, as 'discovery', or its linguistic equivalent, only existed in Portuguese among the European languages (and even there was a recent introduction). The idea of searching the world with the intention of making discovery was the hallmark of this new era in science. Previously, the tendency was not to look outwards but inwards, relying on philosophical musings, and backwards, attempting to interpret ancient wisdom; the Renaissance brought the urge to discover and think anew.

The assumption made by Columbus, based on the science of the time which suggested that there could only be a single landmass on the Earth, was that it would be possible to sail west from Europe to China. That discovery was the result of a misunderstanding based on physics. The main physical theory of the day, developed by the ancient Greek philosopher Aristotle, had everything made of combinations of earth, water, air, and fire. Each of these 'elements' in turn, from earth to fire, had less of a tendency to head for the centre of the universe (assumed to be the Earth). So the broad structure of the world was thought to be a sphere of earth, surrounded by a sphere of water, surrounded by a sphere of air, surrounded by a sphere of fire.

Clearly, this couldn't be exactly true, or there would be no land sticking out of the sea – it would be totally covered in water. So, it was assumed that the earth sphere was off-centre, allowing bits of its surface to rise above the waters. However, if there were just a single landmass, it should be possible to sail in either direction to get to anywhere on the planet – hence the assumption that Columbus could reach China by sailing west.

More recently, accidental discoveries have come thick and fast. For example, the artificial sweeteners saccharin and aspartame were both discovered when a chemist accidentally tasted them while attempting to make something completely different. The non-stick material PTFE was developed when a material spontaneously degraded. And the wonder atom-thick material graphene – ultra-strong and soon to transform electronics – was first produced from sticky tape, recovered from the laboratory bin after it been used to clean graphite blocks.

Famously, penicillin was discovered accidentally when a culture of a virulent bacterium was exposed to mould spores from a window left open by error in the laboratory, which killed the disease. Radioactivity, synthetic dyes and safety glass similarly all came about by mishap. In the last case, chemist Edouard Benedictus accidentally dropped a glass flask containing a plastic substance – the cellulose nitrate held the broken glass together, inspiring the invention.

More mundanely, the ice-cream cone came about thanks to a chance occurrence. Ice cream used to be sold in small cups or glasses, but these were often broken by the customers or not returned. Paper versions started to take over, but at the 1904 World Fair a vendor ran out of cups and improvised by borrowing rolled-up waffles from the adjacent stand (it was hot, and waffles were not selling well) to use as an edible container for his ice cream.

SEEING THINGS DIFFERENTLY

Other misunderstandings led to everything from the invention of the gas balloon to radio astronomy. But accidental inventions aren't all about getting things wrong. In some cases, it's a matter of taking something that was intended to do one thing and discovering that it's far better at something else. This was the case, for example, with super glue – originally a failed material for making clear gun sights, and anaesthetics, where one of the earliest, nitrous oxide, was a failed treatment for tuberculosis.*

A good example of inventive repurposing comes in the invention of plastic film. Since the 2010s, plastic has been increasingly demonised, because of its impact on the environment. And there's no doubt that we had too many single-use plastic items which could (if their users were careless) end up in the ocean. The majority of plastic waste in the ocean is from the fishing industry, and the vast bulk of all plastic waste enters the seas from a handful of rivers in the Far East, so cutting back in the

* Another medical example is the drug Viagra, which was a repurposing of a failed angina medication.

UK, say, has very limited impact – but we need to think more about the environmental impact of the products we use. Having said all that, plastics have done a huge amount of good in keeping food safe to eat, in medical applications and in a vast range of other uses.

One of the earliest ways that plastic came under fire was in the plastic bag. In many countries now, the 'bag for life' has replaced single-use bags. Yet the plastic film that makes up bags, food wrapping and more has been invaluable in improving hygiene and convenience. Once plastic was developed, it might seem obvious to think of producing a wrapping material based on this flexible, protective material – yet the first plastic film, cellophane, had an unexpected change of use.

The Swiss chemist Jacques Brandenberger had no intention of devising a wrapping material when he invented cellophane. After a meal out, when wine was spilled on the tablecloth, it seemed to Brandenberger that restaurants would benefit from owning tablecloths that repelled liquids. This was in 1900, before the development of modern plastics, but there were already water-resistant materials based on the plant compound cellulose.

This common substance – made up of long chains of the sugar glucose – gives structure to the cell walls of plants and is one of the prime ingredients of paper. By the time of Brandenberger's spillage, cellulose was being used to make celluloid (initially as a replacement for expensive ivory in billiard balls and then for photographic purposes) and the artificial fibre viscose. Brandenberger sprayed an orange-coloured viscose solution onto some test material, adding chemical reagents that would convert it back to a layer of cellulose. There was no doubt it would have kept the wine out of the cloth, but there were two problems.

Firstly, the material made the tablecloth too stiff and, more importantly, the plastic did not adhere to the fibres of the cloth – it peeled away as a separate clear film. At first sight, this was a disaster; it was worse than useless. But Brandenberger realised that the film was potentially useful in its own right. It took around eight years for him to perfect his invention and several more to make it commercial – the necessary leap proved to be adding glycerine to make the material more flexible, along with devising a machine to produce it. Brandenberger called his wonder product 'cellophane', combining its cellulose origins with the Greek ending '-phane' implying brightness and transparency. With its ability to keep out bacteria, oxygen and moisture,* cellophane became central to the safe distribution of cuts of meat and other foodstuffs. Its first food use in the US was to wrap the chocolates of the Whitman Candy Company.

It has been argued that cellophane was directly responsible for the development of the supermarket. A butcher, for example, doesn't need cellophane packaging for meat in his or her display. But to allow self-service from a supermarket fridge requires a wrapping that allows the purchaser to see just what they're getting – and to avoid the food discolouring through exposure to the air. The red colouration of much fresh meat comes from the protein myoglobin, which rapidly deteriorates when exposed to oxygen, taking on a brown colour. This doesn't mean the meat has gone bad, but we associate freshness in meat with a vivid red colour. Cellophane wrap made it possible to keep food looking good. The importance of the visual aspect even stretched to sweet wrappers. Such was the impact of cellophane that a 1930 US advert ran 'Your EYES can TASTE Cellophane-wrapped candy'.

Cellophane is still in use today, both for transparent gift-wrap and as the base material for adhesive tape. In fact, it is showing a resurgence for food packaging, where it was largely replaced by cheaper petroleum-based plastics as, unlike its replacements, it is biodegradable. If Brandenberger had succeeded making his wine-repelling tablecloth, the chances are that his invention would have been long consigned to the dustbin of history, but by seeing a better way to use the outcome of

* The original cellophane did hold liquid water, but let water vapour through: a fully waterproof version was patented by DuPont in the US in 1927.

his failed experiment, he was able to give the world a new level of food hygiene.

One final example that is irresistible as it underlines the contribution human stubbornness can make to invention is the potato crisp (chip). This came about when a customer in an American restaurant complained about the thickness and sogginess of the fried potatoes. The chef decided to have his revenge on the customer by producing ridiculously thin slices, fried so long they became rigid, which he then over-salted to make them even less edible. The result proved delicious.

IT'S MOSTLY GOOD

In an earlier chapter we looked at climate change – and it is, without doubt, an extremely serious problem caused by our move away from our purely biological past. But we also need to avoid the hair-shirted, anti-technology mindset that some adopt in response. Taken as a whole, the majority of us live far better lives than our ancestors and – despite the news reports – things are getting better. In medical terms, in the opportunity to make something of our lives, in access to information, and more, the modern 'you' has a better life than your ancestors. Which makes it strange that most of us have a particularly gloomy view of modern life.

The majority of people think that things are far worse than they really are. The late Hans Rosling, a Swedish medical doctor who specialised in the presentation of statistics, carried out widespread surveys around the world on a number of key issues and found uniformly that far fewer people got the answers to questions right about, for example, the levels of world poverty than would be the case if they had simply answered the multiple choice questions randomly. They were strongly biased towards a pessimistic answer.

For example, when asked if in the last 20 years the amount of the world population in extreme poverty has almost doubled, remained about the same, or almost halved, in most countries less than 10 per cent of respondents got the answer right. (In the UK around 9 per cent knew it almost halved, while in the US it was around 5 per cent.) People proved equally pessimistically wrong in questions about girls' education, life expectancy, population growth, natural disasters, vaccination and the level of income in their own countries.

It's hard not to lay at least some of the blame at the foot of the media. The news media love nothing more than a bad story. You will never see a story about this year's great harvest in an African country, only about drought and starvation. It inevitably makes the news when religious fundamentalists prevent girls from attending school – but not when girls are being educated as a matter of course.

We also must consider the contribution of information from charities and aid agencies. Fundraisers stress the problems and disasters faced in the world, and the result is that we give more generously. By distorting the picture, there is a danger that the useful funding that is provided by charities and government agencies does not go to where it can do the most good. When there's an immediate emergency and we need to respond to it, pessimism-driven publicity is essential – but for effective long-term development aid we need a more nuanced narrative.

Rosling was of the opinion that, although the media and charities have some influence on our thinking, this isn't all their fault – in fact, mostly he put our misreading of reality down to what makes you *you*. When he first discovered the gap between people's ideas of the world and what was really happening out there, Rosling thought that the answer would be simply a matter of educating them with the facts – but he discovered that information seemed to have surprisingly little impact. Instead, he believed that our own natural defences get the better of us.

The fact is that you are inwardly programmed (as is everyone else) to try to keep yourself safe and well. This is not exactly a controversial situation. But in the world that *Homo sapiens* lived in for most of our species' existence, this meant reacting to the outside world in ways that have now become counterproductive. For example, we all crave sugar and fat, because they are great ways to keep us going when food is scarce. But when calories are plentiful and it is so easy to pick up junk food, even though we know perfectly well that those ingredients are not good for us, we still crave them.

There's a visual clue to this in the advertising presentation of hamburgers. Photographs of hamburgers designed to entice us almost always show a whole stack of food – not just a burger and a bun, but several meat patties, lashings of melted cheese and sauce, onions, salad – a veritable mountain of calories. Our conscious brains know that such a monstrosity would have more calories in it than we need to consume in a whole day. But something older, a remnant of a food-scarce past, triggers desire. A hamburger is supposed to be a sandwich, but no one could fit one of these things into their mouth.

Similarly, Rosling suggested, we have an automatic response to potential hazards. It's why we can jump at the threat of danger when there is nothing there, or can conjure up a bogeyman out of shadows. We are ready and expectant for trouble. Rosling suggested that just as we crave those fattening foods, we crave drama, because it stimulates the quick decisions we need to get us out of danger. Slow, careful consideration doesn't hack it when there's a predator heading towards you at high speed. Drama gives us that same hit as the fatty, sugary foods. Which is fine if you're dealing with a novel or the movies (or if you are really in danger) – but is something that we need to be wary of when it comes to the presentation and absorption of facts.

THE GREAT DIVIDE (OR NOT)

In assessing how we view the world, Rosling suggests that there are a number of major misconceptions that drive our acceptance of misinformation. For more on this, you can find the details of his book in the Further Reading section – but one thing that is important to pick up on here in helping understand what makes you tick is the illusion of the great divide.

We tend to have a picture of the world that splits it into two: 'haves and have-nots'. Or 'privileged and underprivileged'. It's a black-and-white view, beloved of politicians and supporters of causes, though oddly it comes in two mirror-image forms. To politicians with a domestic agenda, the 'haves' are the minority 'them' and the 'have-nots' are the majority 'us', giving us a wrong that needs to be righted in a fair society. When we're dealing with world issues such as poverty, though, the picture is inverted. It's much more a matter of 'us' being the relatively small, privileged, developed world and 'them' being the sprawling, underprivileged developing world – those who are in need of our help.

It's even possible to display statistics showing a clear split like this, as Rosling did in his book – but sneakily he did this by using data from the 1960s. If you then replace this with current data, you find that the vast majority of the world's population have moved into what would have been considered the 'developed' half, spread out towards the few remaining countries that still fall into the old 'developing' division. There really isn't a clear 'us' and 'them' anymore, we are part of a continuum.

It should be stressed that Rosling was not suggesting that everything in the world is fair and there is no need for aid, but rather that this developed/developing division is an artificial one that does not accurately represent the world we live in. In the more realistic continuum, a good 75 per cent of the world's population sit somewhere in the middle. If we're talking about wealth, for example, they are neither in abject poverty nor in possession of outstanding riches. Of course, that's not to say that there aren't people at both extremes – but they are a relative minority.

In response to Rosling's insistence on discounting terms such as 'developing world', he was often asked 'What should we call them, then?' After pointing out that the use of 'we' and 'them' in that question was simply reinforcing the problem, he suggested a four-way split. For this he envisaged around a billion people currently at level one, living on around $1 a day, 3 billion on level two, earning around $4 a day, another 2 billion on level three, reaching $16 a day, and a remaining billion at level four on $32 or more a day.

In practice, the chances are, if you're reading this book, that you will be on either level three or, more likely, level four. And if on level four, it is true that you are in a relatively rich minority. But you need to be careful not to categorise the rest of the world as a 'them' for whom life is disastrous. And we also need to recognise how much things have moved on. As Rosling pointed out, 200 years ago, 85 per cent of the world's population was on level one. In the 1950s, Europe and the US were almost entirely in levels two and three. Now the percentage of the world in level one has fallen to around 14 per cent, while the majority of Europe and the US have shifted up to level four.

BUCKING THE TREND

It's important to make clear again that there is no suggestion here that things couldn't be better than they are – they can, and hopefully will be. The fact remains, though, that you are likely to have an inaccurate picture of the world. Mostly you are likely to have misapprehensions about how bad things are (or, rather, aren't) at the moment. However, there are some exceptions which need highlighting. We may, on the whole take a pessimistic view, but there are few cases where our perception tends to be rosier than reality.

In his book *Perils of Perception*, Bobby Duffy, Professor of Public Policy and Director of the Policy Institute at King's College London, echoes many of Rosling's findings, based on a series of large online polls taken across 40 countries. But there are a number of cases where the results tend to underestimate the negative. One that Rosling also pointed out was on the subject of climate change. Rosling went for a simple 'experts expect it will get warmer/colder/stay the same', and not entirely surprisingly, the correct 'warmer' came out on top everywhere (percentages ranging from the mid-70s to the mid-90s).

Rosling put this down, at least in part, to the approach taken by former US Vice President Al Gore using fear and exaggeration to spread the message – making the message stronger, but likely to distort it as well. This is the approach taken to the extreme in 2019 by the organisation Extinction Rebellion and the activist Greta Thunberg. As Rosling points out, the danger of using fear and exaggeration is that when the extreme outcome doesn't occur soon, as is likely to be the case most of the time, the whole message is damaged.

However, when Duffy took a more detailed approach, our perception of climate change is not so accurate. For example, when people were asked in 2018 to guess how many of the hottest years on record occurred within the last eighteen years, on average they answered nine. The correct answer is the far more significant seventeen. It seems likely that Rosling's relatively accurate results were in part due to the general nature of the question, but also because he asked about the opinion of experts. It's likely that quite a few of those who deny the reality of climate change would agree that experts expect warming – but they don't trust the experts.

Another example of a tendency to rose-tinted spectacles comes in the US attitude to gun crime. Duffy shows that around 80 per cent of those who support the Democratic Party (and hence are more likely to support gun controls) believe that more Americans are killed with guns than knives or other violence, while only 27 per cent of core Republican supporters believe this to be the case. The fact is true, implying that for the Republicans polled, perception was strongly influenced by their desire for a particular outcome, one that they saw as more positive than did their Democrat rivals.

It seems – not entirely surprisingly – that topics with a strong emotional content are more likely to produce incorrect perceptions, and these are perceptions which will continue to be held despite contrary evidence. For example, in most countries, people think that the number of immigrants in the population, often an emotionally loaded topic, is considerably greater than it really is. In the UK, for example, Duffy's surveys produced an average guess of immigrant numbers at 25 per cent where they were actually at 13 per cent. (The US came up with 33 compared with a real value of 14, and Australia 38 compared with a real 28.) When those questioned in some countries were asked a follow-up question after being showed the actual numbers, they were more likely to believe their own estimate than the official figures.

POLLS AREN'T FACTS

There is some danger in relying on polls to give an accurate picture of public views. Sometimes this is because an apparently simple question can be difficult to answer 'correctly' when based on a complex set of data. For example, Duffy reminds us of the mistaken attempts of UK politicians Boris Johnson and David Cameron to estimate the prices of everyday items in 2013, which caused great hilarity in the press. When Cameron said a loaf of bread costs 'well north of £1', an interviewer corrected him to 47p. Duffy states that the interviewer was 'very wrong, it was more like £1.20 at the time'. Yet Duffy's figure is only correct for an average between artisanal bread and supermarket loafs. The 47p figure was approximately correct for a mass-market basic loaf.

Similarly, Duffy criticises respondents to a survey on the price of a pint of milk for saying it was '29p or less', where Duffy claims the real figure in 2013 was closer to 49p. This may have been true when buying a single pint bottle; however, going on the bottles stocked on supermarket shelves, far more people buy milk in 4-pint bottles, which would have made a guess in the 25–30p per pint range correct. These errors seem to be primarily the result of not making the question clear. But it is also sometimes the case that the figures used to make an argument are simply wrong.

A good example of this is a country's GDP figures.* Regularly in the UK (and many other countries) we hear of how GDP in an earlier quarter has gone up or down. The most recent such announcement at the time of writing was in August 2019 when we were told that GDP between April and June of that year had fallen by 0.2 per cent – and the media was full of speculation of how the threat of Brexit has caused this fall. Yet, as British journalist Michael Blastland points out in his book *The Hidden Half*, the official GDP figure is repeatedly corrected over the months that follow as better data becomes available. The GDP value can shift by as much as 1 per cent, and typically changes from the first published value by about 0.4 per cent. Drawing any conclusions from a 0.2 per cent fall (or rise) on the initial headline GDP is totally meaningless.

We have to accept that any value where we don't have total data (or, in the case of a forecast, have no data at all) is likely to include an error, the level of which is usually estimated in science, but which is rarely mentioned when statistics are presented to the public. And there is also the potential for confusion over the way that statistical data is presented. Duffy points out that in the UK, both before and after the 2016 referendum on leaving the European Union, British citizens were likely to overestimate the levels of EU immigration. However, what Duffy didn't know when he wrote this was that the UK government had significantly *underestimated* levels of EU immigration. In 2019, the UK government's Office for National Statistics (ONS) pointed out that they had understated immigration from the EU by around 16 per cent. In Duffy's survey, taken before the error was revealed, the public overestimated the level of immigration by 10 per cent.

* GDP or 'gross domestic product' is an attempt to measure the value of the goods and services produced by a country. It is now considered to be a highly flawed measure. It was designed primarily to deal with traditional manufacturing and agriculture and is an uncomfortable fit to the now extremely important service industries and fails to properly account for the impact of innovation.

There are two significant points in the previous paragraph. One is that the official figures (which came from a survey, not direct data) were highly inaccurate. The second is that the old saying 'there are lies, damned lies, and statistics' isn't entirely spurious.* While the final two sentences of the previous paragraph are technically accurate, the apparent conclusion – that with the error fixed the public *underestimated* EU immigration by 6 per cent is totally wrong. This is because the ONS correction was on the figure that politicians usually bandy around: the net immigration for the year – how many extra people from the EU were added to the population. But the question in Duffy's survey was about the percentage of UK residents who were EU nationals – it wasn't about immigration rates at all. This is a much bigger number. The ONS underestimated annual EU immigration by an average of 35,000 a year, but the public overestimated total EU residents by over 6 million.

Like Rosling, Duffy looks for reasons for the difference between reality and perception and how this varies around the world. He agrees that we have a strong inbuilt protective bias to negativity, but suggests that cultures which put a strong emphasis on emotional content to argument (such as Italy and the US) tend to exaggerate values to make their point, while cultures that tend to be unemotional in their arguments (such as Sweden and Germany) tend to have perception that is closer to reality. As we have seen, though, correlation is not necessarily causality, and though there seems a kind of logic to a tendency to exaggerate if you are imbuing an argument with emotion, the causal link has not been proved.

What shouldn't be in any doubt at all, though, is how much better things are for human beings in most of the world than they were prior

* The phrase was made popular by American writer Mark Twain, who attributed it to the indubitably witty British prime minister (and fiction author) Benjamin Disraeli. But no one has been able to find it in Disraeli's output.

to the contributions of human ingenuity. I can't stress enough how much a significant part of what makes you *you* is the impact of these developments. But we can't dismiss biology, and it is important to return to this in what is probably the best-known debate over what shapes you as a person: is it nature or nurture?

9

THEY DON'T MESS YOU UP, YOUR MUM AND DAD

'It's in their DNA' has become a cliché to refer to a behaviour or aspect of a person that is inherited, as if each of us is run by a computer program that uses DNA as its code. DNA, the molecule that specifies our genes (and far more) is a wonderfully flexible structure. When we speak of a chemical compound, it usually has a specific formula. So, familiarly, common salt is sodium chloride, with the chemical formula NaCl, or benzine is a hexagonal ring of six carbon atoms, each with an attached hydrogen atom, represented by the formula C_6H_6. But I can't tell you the chemical formula for DNA – or deoxyribonucleic acid to give the compound its Sunday best name. That's because it is not a specific substance, it's a chemical framework used to store information.

IT'S ALL IN THE DOUBLE HELIX

Your DNA comes in the form of chromosomes – the majority of your cells contain 23 pairs of such chromosomes, each of which is a single molecule of DNA. The biggest of these, chromosome 1, contains around 10 billion atoms – that's chunky. Most of us are familiar with the basic appearance of DNA – two corkscrewing twisty bits (known more technically as a double helix) linked by a series of straight segments, rather like the steps in a spiral staircase. The twisty bits are polymers – long chains of sugar molecules connected together by smaller phosphate components. The sugar in question is ribose, which is in the same group of compounds as the more familiar sugars glucose and fructose.

These polymers are important to provide structure – but it's the 'tread' parts of the spiral staircase that make DNA so valuable to all living things. Each tread is made up of a connected pair of compounds called bases, of which there are four different types: cytosine, guanine, adenine and thymine. The clever part is that each base only ever pairs up with a specific one of the others – adenine with thymine and cytosine with guanine.* This standard coupling regime gives DNA a mechanism to be able to split into two and create two copies of itself – essential to support the mechanism by which organisms grow, where cells divide into two.

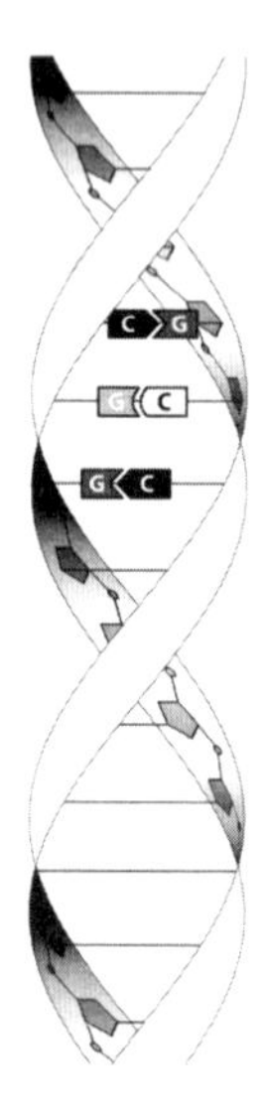

A section of DNA, showing the pairing of bases.

* If you want to impress friends and relations by remembering which base forms a pair with which, think of the bases represented by their first letters: C, G, A and T. The straight-line letters A and T pair up, as do the curvy letters C and G.

As a result of the pairing, DNA can be unzipped down the middle of each base pair. It can then be reconstructed by adding the appropriate pairing base as the new piece of DNA is constructed. Why does this matter? Because DNA is not just a molecule, it's a data store. Just as a computer stores its information in binary zeros and ones, a biological organism stores its information in quaternary – holding one of four different values depending on the base that has been deployed.

The big breakthrough in understanding DNA came in the 1950s. Its existence had been known for around 100 years, but it was in 1953 that this sophisticated structure was discovered. In a computer, bits are usually lumped together in 8-bit chunks called bytes. In DNA, groups of three bases, called codons, are used to indicate which amino acid building block is going to be used next when constructing the vitally important biological molecules called proteins. In principle there could be 64 different amino acids specified by these groups of three, but in practice, nature duplicates the codes considerably, so the codons only specify a choice of 20 amino acids, plus special 'start' and 'stop' codons that indicate how to chunk together the set of amino acids to make a protein.

WORK OF GENIUS

The word we tend to associate most with DNA is probably 'genes'. You can't look at a piece of DNA (however good your microscope) and see genes. A gene is a collection of bases in a stretch of DNA that has a function; it's not a physically distinct object. In this respect the DNA is a bit like memory on a computer. It's only if you have the right 'map' that you can find information in computer memory, and similarly, to use a gene requires extra information to know how to process it. A typical gene might consist of the series of codons that specify a protein, plus additional bits of DNA that control how those codons are used.

To make things even more complicated, the DNA itself isn't used directly to produce proteins, but rather the information from it is copied to a simpler equivalent of DNA called RNA, which is then used as the template for producing the protein (as well as in other roles). This process is undertaken by tiny but extremely complex 'molecular machines' known as ribosomes. 'Tiny' is, if anything, an understatement. The cells of your body are small enough to need a microscope to see them, but each cell in your liver, for example, contains around 13 million ribosomes. And a ribosome itself is a fearsomely intricate mechanism that acts like a construction machine.

If that seems non-trivial, there's even more required to make this process work. This is because (unlike the genes in bacteria) our genes aren't neatly laid out in a contiguous row. Instead, they are split into pieces in the DNA with often long interposed stretches known as introns between these component parts. All of this is passed over to the RNA when making use of genes. Before the ribosomes get to work, the molecular machinery in the cell has to cut the introns out of the RNA and splice the gene together before it can be used. It's a bit like watching a TV show that has adverts using a clever personal video recorder that omits the boring bits.

This DNA business may seem unnecessarily complex, but DNA is present in every living thing on Earth. We all share this essential information mechanism to help with the construction and functioning of our organism. It ought to be stressed, though, that DNA is not all about the genes. In fact, only about 2 per cent of your DNA forms genes. Much of the rest was once regarded as 'junk', amassed but not used over the millennia of our evolution. And certainly, some parts do appear to contain unnecessary repeats, perhaps from a copying error long ago. But we now know that much of our DNA outside of the genes has functions, including instructions for switching genes on and off in response to environmental triggers.

Without doubt, your DNA is crucial to making you *you*, and we need to revisit our early venture into family trees to complete the picture, because your DNA comes from your parents. Before the role and structure of DNA was understood, it was already realised that we get influences on our nature from both our parents. But DNA shows us how this happens.

Remember that you have two sets of chromosomes, one of each pair from each of your parents. But, of course, each of your parents also had two sets of chromosomes. You don't get handed one of their sets wholesale – in the process of producing egg and sperm cells, your parents' chromosomes go through a process called genetic recombination where sections are brought together from each of their versions, producing a whole new combination – this is why you may resemble your parents in some ways, but you are still a whole new, unique entity. There are also some errors that occur in the process of assembling your DNA, meaning that you (like all the rest of us) are a mutation. A relatively small amount of your genetic material will differ from that of both your parents. Typically, you will have around 4 million single-letter mutations in the base pairs in your DNA.

DON'T FORGET YOUR FELLOW TRAVELLERS

We are going to concern ourselves here with your DNA, but it's worth remembering that this isn't the only DNA you have in your body – far from it. Your body is home to trillions of bacteria and other microorganisms, collectively known as your microbiome. When I wrote *The Universe Inside You*, the widely accepted figure for the number of these single-celled organisms was around ten times the 10 trillion cells in your body – it's now accepted that's it's more like the same number of cells. But that's still 10 trillion other organisms to one of you.

Clearly your microbiome has an impact on your health. Many of your bacteria are friendly and help out, for example, with digestive processes, though others cause various unpleasant diseases. But, perhaps surprisingly, there is some evidence that there can also be an interplay between your genetic makeup and your microbiome. It's always difficult to be absolutely sure where there is a causal link between two aspects of your body, but there seems reasonably good evidence that some aspects of your genetic makeup will influence the bacteria that are more likely to turn up in your gut, and hence have an influence on whether or not you suffer from digestive problems, or are at risk of certain diseases.

IS IT ALL IN THE GENES?

Whether or not one of its routes to influence is the microbiome, there's no doubt, as we shall discover, that genetics has a significant impact on what you are as an individual. Admittedly we only differ from other humans in around 1 per cent of our DNA, but that is enough to make significant differences in appearance, so why not also in our personalities? Long before Charles Darwin's half-cousin Francis Galton come up with the phrase 'nature versus nurture' in 1874 there was already a debate on just how much impact what we now know to be genetics has. In the old days, when the only tool available to assess the 'nature' side was the family tree (see Chapter 2), there were serious problems. Just because you are descended from someone it doesn't mean that you have their genes.

This may seem counterintuitive. You certainly have a mix of your parents' genes, and they of their parent's genes (and so on). But remember how the real, genetic family tree opens up, branches and intermingles. At each generation the contribution that comes from one individual is diluted, to the extent that, going back far enough, there may be none left. So although, as we have seen, it's pretty much certain should you be European in origin that you have the Emperor Charlemagne in your family tree (see page 21),* it's entirely possible that you don't have any of his DNA.

* Remember, though, that if the Emperor Charlemagne is one your ancestors, so is everyone alive at the same time who has living descendants. Charlemagne himself is

Back in 2001, a major scientific effort called the human genome project announced (somewhat inaccurately) that it had completed the task of deciphering the human genome, where a genome is the total collection of information in the DNA of a particular set of chromosomes. I remember my copy of the leading scientific journal *Nature* coming with a pull-out wallchart, as if this were the culmination of a football competition, rather than a serious scientific endeavour.

At the time, it seemed as if this scientific effort had produced the key to seeing what made an individual what they were. Scan through a genome and surely you would spot the genes for all the characteristics that gave that person his or her individuality. Here's the gene for a good sense of humour. There's red hair, and there for the tendency to grind your teeth. Oh, and over in the corner, is that the gene for being a Manchester United fan?

Except it's not like that. After the initial euphoria, the limitations of being able to map out a person's genome were realised. Although there are a few diseases, for example, that can be pinned down to a single faulty gene, the vast majority can't. Nor can many other individual characteristics be tied to down so easily. It's true that, for example, red hair is the result of a mutation in a single gene. It's a relatively rare one, occurring in under 5 per cent of the world's population as a whole, though in some communities – the UK provides an obvious example – it can be a lot higher.

Red hair is not caused by a single mutation – there are several ways the gene can be changed to produce a variant on the pigment melanin responsible for hair colour – but it is the responsibility of a single gene. More often than not, though, there are a whole host of genetic contributions that combine to make up any particular characteristic of 'you'. We always like to think we can spot aspects of appearance or character

....

not special here, he's just a useful peg. (And he has a cool-sounding name compared with, say, Pepin the Short.)

that are passed on from one parent to a child, producing such remarks as, 'He's got your chin!' However, parents of adoptive children will tell you that they frequently receive exactly the same comments, although there is no genetic connection whatsoever. We see similarities that we want to see.

Not only do genes come together in all sorts of interesting combinations to produce a specific outcome, but bear in mind, as we have seen, that 98 per cent of our DNA isn't made up of genes. And it has been discovered that so-called epigenetic factors – when, for example, different genes are enabled and disabled – have just as much importance in respect of what makes you *you* as which genes are doing what.

It's not that genomes haven't proved useful. As we will discover, genetic research has enabled us to discover a huge amount about what does and doesn't contribute to our individuality. And a complete genome is now far more easily deciphered than was the case with that original multi-billion-dollar, multi-year project. But, as is so often the case with science, we have to be aware of the proviso that 'It's more complicated than we first thought'.

THE SURPRISING BALANCE OF NATURE AND NURTURE

What about the nurture bit, though? How are you influenced by what you've experienced in your past, whether it's being brought up in a loving family or suffering a childhood of deprivation, whether you can be considered privileged, deprived or normal (whatever that is)? There is no doubt that our environment has an impact on us, but surprisingly, the obvious contributors such as family upbringing and education are unlikely to have as much effect as was (and often still is) assumed to be the case.

It is partly as a result of convention that we give a lot of weight to nurture and immediately look to someone's upbringing as something that shapes their personality and behaviour: as the discipline of psychology developed from early, often loosely substantiated, ideas, it was taken as a given that it was the environment, particularly the childhood environment, that shaped our personality. Early, unscientific disciplines such as psychotherapy were built on pure assumption, rather than scientific evidence. And when scientific evidence began to be properly assessed, its message proved to be quite a shock as it showed that those early assumptions were far from the truth.

In a remarkable long-term study, two American sociologists, John Laub and Robert Sampson compared data on a range of men from similarly deprived backgrounds who had suffered childhood poverty* and had engaged in criminal activities while still children. The researchers were able to follow these individuals through to the age of 70, because they discovered 40-year-old records from a study filed away at Harvard Law School and subsequently tracked down many of the participants of that original trial to see how their lives had turned out.

The striking outcome of the research was that there was no way to categorise the boys based on their original circumstances that would indicate how they would develop and act through their lives. The way they were brought up, poverty, mistreatment, poor performance in school – none of these proved an effective indicator of how a particular individual would behave in later life, when some would go on to have fruitful and positive adult lives and others would become persistent serious offenders.

It's not that their difficult beginnings had no impact. The participants in the study were more likely to have difficult adulthoods than those from a different background. But the outcome for any individual was not predictable from any of the 'obvious' contributory factors in their early years. Undoubtedly a significant part of the outcome was down to random circumstances. Luck, or lack of it. Where the boys went, whom they met – but the scientists were unable to find any pattern

* This was absolute poverty – lacking basic needs like food and shelter – rather than the definition often used now which is relative poverty, which involves falling below a selected percentage of average income in a specific country.

in what would previously have been assumed to be the causes of their behaviour.

The broad feeling now, backed up by a lot of research, is that about half of what makes us what we are is genetic and about half is environmental. In each case, things are more complicated than was once thought. As we have seen, on the genetic side, it's very rare that anything can be pinned down to a small number of specific genes. Often there are hundreds of interacting genetic influences. And at the same time, the environmental side only features small contributions from the obvious influences such as parents and education. Again, there are so many small interacting factors that shape our lives, which make it very difficult to say exactly how a particular environment will influence any one individual.

GENES VERSUS ENVIRONMENT

The researcher who has arguably done most to quantify and clarify the whole nature versus nurture debate is American behavioural geneticist Robert Plomin, based at King's College, London.

Plomin comes down solidly on the side of genetics. He opens his book on the subject of behavioural genetics by saying: 'What would you think if you heard about a new fortune-telling device that is touted to predict psychological traits like depression, schizophrenia and school achievement? What's more it can tell your fortune from the moment of your birth, it is completely reliable and unbiased …' What he is referring to is your genetic makeup. In a sense what he says is true, but the way it's phrased is also misleading. As mentioned above, your genes only contribute about half of the factors influencing your psychological traits. And we often don't know what exactly it is that is contributing the genetic factor.

We always have to be careful when looking for causes when dealing with a complex, messy system like a human being. A lot of the studies that generate news headlines show this. To be able to be sure that a factor is responsible for producing a particular effect, the good scientific approach would be to run repeated large-scale trials, keeping everything else the same but varying just that factor. However, it's just not possible to run these kinds of trials on human beings for both ethical and practical reasons. Instead, more often than not, it's necessary to go with data from the real, messy world and to try somehow to counter the impact of every other factor.

So, for example, when you read that the Mediterranean diet reduces the risk of heart disease, that diet is simply a tiny part of the environment experienced by the people in this study. They may also be more likely to live near the sea or may live less stressful lives than city dwellers. They could favour a particular pattern of alcohol consumption or take more exercise than others. And so on, for hundreds of other possible influences. Those undertaking such research have to manipulate the data to deal with other factors – for example, trying to remove the impact of risk factors such as smoking, air pollution and stress. This is very difficult to do with any accuracy.

Where possible, researchers make use of natural situations which produce the kind of controls they would like to impose in an experiment. So, for example, if we were trying to discover the relative impact of genetics and environment, we would want to split off people into groups with the same genetics and different environment, and similarly to look at those who share an environment but have different genetics. To a degree, this can be achieved by devising studies using twins and adoptees.

Identical twins are clones: they are (almost) genetically identical and are usually brought up in the same environment, though a small number are brought up separately, allowing for 'same genetics, different environment' studies. Non-identical twins, meanwhile, who are siblings born together but not clones, are not genetically identical but are usually raised in the same environment as each other. Adoptees are genetically different to their adoptive parents but share an environment with them. Where there is more than one adopted child in the family, then the adopted children can be compared with each other, as their environmental influences will be more similar than comparing them with their adopted parents.

These types of studies are common in psychology, making it possible to attempt to distinguish 'nature' factors and those that come from nurture. As we discovered on page 24, even twins are not totally identical genetically, and starting with twins in a shared environment does not result in identical consequences. Similarly, adoptees can produce misleading results as two siblings (adopted or not) still have different environments in the outside world. Even in the home, each sibling has the other's potentially very different behaviour as part of their environment. Nonetheless, these studies do provide valuable information in attempting to assess how the influence of nature and nurture comes together to produce you (or anyone else).

Robert Plomin was involved in both a major adoption study in the US, using 250 adoptive families and 250 control families,* and a huge twin study in the UK, with 16,000 initial participating families. These studies were extended over decades to collect data on the development of the individuals – in the case of the twin study, this involved contacting the families regularly up to the twins being age 21 – the result has been the collection of over 55 million items of data. These studies and others give us a good feel for the balance between nature and nurture for a whole range of the psychological factors that help make you *you*.

* In scientific studies, a control is a way of comparing the results from the subjects you are interested in with similar subjects lacking the same feature. So, for example, when testing a new drug, a control would be a placebo – something that is apparently a drug but has no active ingredients. In the adoption study, the control would be families with a similar structure and circumstances, but where the children are not adopted.

Broadly speaking, as we have seen, these studies confirm that your psychological makeup is around 50 per cent down to your genes and about 50 per cent down to your environment. That's a 50:50 split between nature and nurture, to employ the usual terms of the debate – except it's worth noting here why 'environment' is the more precise designation. Children brought up in the same home environment differ from each other by a similar amount to the general population. They have that 50 per cent genetic contribution, but the rest of their makeup seems pretty much entirely down to complex interactions with the broader environment. The only significant exceptions seem to be in religious and political beliefs, which do have significant initial home influence, but even this wanes as we mature.

LIVING IN CHAOS

The trouble is that the environment each of us was brought up in was chaotic. As we have already seen, this is not in the general sense of the word, meaning pure disorder, but rather reflects the mathematical concept of chaos. Chaos theory tells us that where systems have the potential for different parts of the system to affect each other over time, the systems are likely to be very sensitive to initial conditions. Extremely small and hard-to-detect differences in their starting environment can result in huge variation down the line. The result is something that can look like randomness, though it is, in fact, deterministic.

Chaotic systems can be complex – the weather, for example, is a complex chaotic system – or surprisingly simple. A pendulum with a hinge part way down it jumps and jerks around in an apparently random fashion, because the interaction between the two segments is sufficient to push it into chaos. It is hardly surprising that the system comprising you and all the influences in your environment is chaotic enough to ensure that many small experiences generate a huge impact down the line. A paper published in 2008 underlines the extent to which microscopic interacting factors can produce huge environmental differences in an organism. The revelation arose as a result of the discovery of something remarkable in its own right – a parthenogenic crayfish.

The crustacean in question is thought to be a variant of the American blue crayfish. The species seems to have first turned up in mysterious circumstances in a fish tank in Germany, where it was given the name Marmorkrebs (marbled crayfish) because of its distinctive grey marbled appearance. But that's not what is remarkable about the animal. It's that 'p' word: parthenogenic. This means that the Marmorkrebs reproduces asexually. A single, isolated marbled crayfish can (and does) produce lots of baby crayfish.

This makes the Marmorkrebs a potential hazard if it reaches the wild, as it makes it far easier for it to take over a new area if accidentally released.* However, the scientists who first studied it realised it also made the animal a great research tool for studying the way that organisms develop with particular characteristics, as each baby crayfish is a clone of its mother – they should be near enough genetically identical.

In an experiment using them, written up in 2008, the German researchers raised batches of the crayfish in what were, as near as possible, identical conditions. They were fed the same way, in the same environment, even looked after by the same individual in case there was an effect from how a particular person treated them. Being clones meant that, mutations apart, the genetic side was unvarying, while every effort was made to keep the environmental side consistent too. Both of the

* It's a reasonable question, if parthenogenic organisms are so good at taking over an environment, why sex is so common in nature. In the short term, parthenogenesis provides rich rewards, but because the offspring lack the mix-and-match nature of sexually produced genes, they are less likely to come up with variants that are able to respond to changes in the environment, so long term tend to be less successful.

key factors were controlled. And yet the outcome was a surprise.* The crayfish grew up to be hugely different from each other.

The mottled shell patterns varied wildly from crayfish to crayfish. The biggest siblings were twenty times bigger than the smallest. The lifespans of the animals varied hugely too, from 437 through to 910 days. Even their behaviour was significantly different, from their way of moving to their social interaction. Clearly something was influencing the individual crayfish to be different.

The changes in pattern were perhaps the least surprising outcome. In 2001, the first cloned cat, known as Cc, proved to be anything but the carbon copy of its mother that its name suggested.† The mother was calico (white with patches of orange and black), while Cc was tabby (brown striped with an M shape on the forehead) and white. This difference was caused by epigenetics, the way that genes are turned on and off by environmental factors and developmental processes. Supposedly, the careful control of the way that the crayfish were raised should have minimised the possibility of this occurring. Yet on measuring one of the epigenetic mechanisms, called DNA methylation,‡ there was significant observed difference between the clones, though the variation could not be matched to the obvious changes in the animals.

Admittedly, the scientists studying the crayfish tried to keep their environments the same, but their attempt surely underestimated the challenge they faced. Chaos theory originated when the American meteorologist Edward Lorenz was working on an early weather forecasting computer program. The computer was very slow, so when re-running the program, instead of starting it from the beginning,

* The cynic in me wonders if the outcome really was a surprise. Why would you undertake such an experiment if you didn't expect something interesting to happen?

† Allegedly, the name stood for Copycat rather than carbon copy.

‡ In DNA methylation, methyl groups (a carbon atom with three hydrogen atoms attached) are bonded onto the outside of the DNA, changing the behaviour of the genes.

Lorenz input values that had been printed out part way through an earlier run. The result was a totally different forecast.

What Lorenz realised was that the computer worked to a higher level of accuracy than the printout had displayed. The machine actually worked to six decimal places, but to save paper, the printout had only shown the first three decimal places of each number. As a result, if the computer had been working on a figure of, say, 0.634152, what was printed out was 0.634. That tiny difference had resulted in a big divergence over time. There was no way that the biologists could control the environment in which the crayfish were raised to make their lives similar in a way that was comparable with Lorenz's full six decimal places – like twin siblings, for instance, their lives were similar but not identical.

It could have been, for example, that one ate more food on a particular day because of its position in a tank when the food was provided. It may have been that the water temperature or the chemicals dissolved in the water varied slightly from place to place in the tank at crucial moments. Even the assumption of identical genetic structure was almost true, rather than exactly true. Clones don't have a 100 per cent identical genetic makeup. The biological processes used to copy DNA are impressive and have error-checking mechanisms, but some errors creep in. Even outside the copying process, minor changes can be introduced, for example, by the impact of cosmic rays, high speed particles from outer space, or background radiation.

In a sense, then, the dramatically different crayfish were not demonstrating anything new. Their makeup was still a combination of genetics (plus epigenetics) and environment – nature and nurture. But the implication is that, if this is a chaotic system, it's much harder to forecast anything from the combination of the two because very subtle variations can have very large impacts. The jury is out – a chaotic system is my explanation, but the researchers simply left it as 'intangible variations' and 'developmental noise'. The fact remains once more that it's more complicated that we thought.

Some argue that this kind of discovery suggests that there is a third factor in our makeup – nature, nurture and circumstance, which is added into the mix to take in the combination of many tiny variations that seem to influence the way the crayfish develop. Yet it's hard to see why this isn't just a refinement of the environmental contribution. Circumstance may not have the seemingly conscious implications of 'nurture' – but that just reflects a label chosen more for its assonance with 'nature' than any suggestion that it's always about actions taken to make something happen, rather than random experience.

GENES AND PATHWAYS

The outcome of studies like Plomin's and the research on the crayfish, then, seems to explain the underlying factor that allows individuals brought up in the same nurturing environment to be significantly more different than their genes alone provide for. Some have interpreted the finding that we can't allocate much influence on our personalities to our home lives as akin to saying that a loving home environment makes no difference – yet it clearly does in the sense that it can make it easier for the individual to deal with what's thrown at them and to be happier in life. These genetic and environmental differences shaping your psychological makeup may not be much influenced by the way you are brought up, but a nurturing family will ensure you have the support to make the best of what genes and environment throw at you.

The same, broadly, also applies to school and life experiences. They will, of course, generate different potential paths in your life. So, for example, it's the case at the moment in the UK that students at private schools are significantly more likely to attend the top universities than students at state schools. Some of this is down to the nature of the schools – far more private schools are selective than are state schools, so you would expect students from the private sector to do better on average. However, some of it is down to being given the opportunity to take a particular path.

Although top-ranked universities such as Cambridge are doing a lot to encourage a wider range of school students to see Cambridge as a potential next step,* private schools put a higher percentage of their students forward to top universities than do state schools. But note that this is not saying that the school has made the student a different person. A school can put you on a particular route, but makes little difference to who you are.

A similar potential misinterpretation of the data arising from the combination of genetic and complex environmental influences would be to deduce from the way that certain groups do less well academically, or are more likely to have a criminal conviction, that these groups are genetically inferior to those who do better. The genetic and environmental combination that makes you who you are does not provide you with opportunities. These come largely from a combination of luck with cultural and social factors. Of course, what you do in response to these opportunities *will* depend on your internal makeup, but it doesn't mean that differences in opportunity don't exist.

GENES CAN DISTORT THE ENVIRONMENT

One remarkable discovery from the research was that even factors that appear to be purely environmental can in reality be in thrall to the genes. A powerful example that Plomin explores is the idea that parenting style has an influence on the behaviour of the growing child. Based on data, rather than 'common sense' hypotheses, Plomin shows that the influence is mostly the other way round. The behaviour of the child influences the parenting style – and the child's behaviour that does this has that strong, circa 50 per cent, genetic source. It produces a kind of behavioural feedback loop.

* In 2019, 68 per cent of students starting at Cambridge were from state schools – this has risen from 52 per cent in 2000. Around 85 per cent of potential applicants attend state schools.

The reason Plomin puts so much emphasis on the genetic side* is that it is much harder to pin down which environmental conditions are providing the other half of the contribution to behaviour – so the genetic half is the bit that can be sensibly quantified and is the largest single factor, even though it may be the result of the interaction of many different genes – but that doesn't mean that the environmental half doesn't exist. Even so, it's a powerful statement that needs further unpacking to help determine what makes you *you*.

When this kind of linkage was first announced, there was a backlash, even among professionals in the field. Plomin showed, for example, that there is a genetic component to how much TV we watch as children. This disconcerted some professionals, as TV watching was one of the standard environmental tropes of the period – something that had been assumed to be purely environmental in nature. Plomin notes that a 'prominent behavioural geneticist' wrote that 'no gene for TV watching, a behavioural phenotype† non-existent three generations ago, could plausibly exist.'

This is a bizarre argument from someone who presumably knew what he was talking about. Of course, there is no 'gene for TV watching' – nor was there any suggestion that such a gene existed. Rather, the sensible inference from the research was that your genetic makeup contributes to your attitude to watching TV. As Plomin puts it 'We can turn the television on or off as we please, but turning it off or leaving it on pleases individuals differently, in part due to genetic factors. Genetics is not a puppeteer pulling our strings. Genetic influences are probabilistic propensities, not predetermined programming.'

There is no suggestion that genetics is the sole cause of what are often seen as environmental factors – it rarely makes up more than 50 per

* Other than being a behavioural geneticist.

† 'Phenotype' is one of those words beloved of academics that seem designed to conceal rather than help communication. Your phenotype is just how you look and act, as opposed to the genotype, your genetic makeup.

cent of the source of an 'environmental' influence of behaviour. Yet it is very frequently the largest individual influencer among the various components that make up the reason for our behaviour. What behavioural genetics tells us about the influences of nature and nurture can seem strongly counterintuitive, however much the data tells us what is really happening. This is not unusual, though, in science, from our lack of intuitive grasp of randomness and probability to the seemingly bewildering nature of quantum physics.

BACK TO THE BEGINNING

We've already seen the surprising input that genes can have on 'environmental' factors influencing our behaviour – and this unexpected relationship comes through even more when we look at the way that the nature/nurture split changes over time. It seems entirely sensible that when we're born, and have had few opportunities for the environmental influences of nurture, that we are constrained by our genetics – our nature. Then, as we gain experiences, we would expect that the environment's influence would get stronger. However, the data shows that reality turns this idea on its head.

When identical twins are compared with fraternal twins (who are no more genetically related than any other siblings), looking at the way that they change as they grow up, the identical twins become more similar behaviourally, while the fraternal twins become less so. The implication is that the genetic contribution to the way they behave grows with time, rather than being overwhelmed by the environment. Perhaps one touch of common sense remaining here is that we're all aware of becoming more set in our ways as we get older – perhaps this reflects an increasing importance of the genetic component.

The reality is a little more complex than the headline effect. Where, for instance, the genetic component of intelligence seems to become increasingly dominant over time, what feels like it should be a related aspect of our nature, school achievement, does not seem to change its nature/nurture balance, sticking around the 60 per cent nature level. Plomin suggests this might be because we teach the skills which are measured as school achievement, whereas we don't do much to teach children the skills that are labelled as intelligence, allowing the genetic influence to dominate over time.

FROM DISORDERS TO SPECTRA

Arguably, the biggest change in the way we need to think about what shapes us from the discoveries of behavioural genetics comes in our attitude to mental illness. The very term 'mental illness' suggests a category error. When we go to the doctor with a physical illness, there is usually a specific cause – a virus, or a bacterium, or a physical irregularity in the body. Either you've got an illness, or you haven't. Through lazy thinking, we tend to take a similar approach to mental illness. We think that it's something you've either got or you haven't got, as if it similarly had a single cause. But while it's true that there are some conditions resulting from a specific physical defect in the brain's mechanism, often what we label mental disorders are the impact of being at one extreme of a large set of possible genetic combinations.

Note that this is not dismissing the severe problems that can arise from such a genetic makeup, but the point is that there is not a simple breakpoint between having a disorder and not having it – we are all on a spectrum of degree of influence of any particular genetic combination. To take a common example, this would mean that dyslexia is not a condition, but rather those we label dyslexic are on a different position of the spectrum of genetic support for reading capability.

With most psychological issues there are large numbers of genetic factors which, depending on their total contribution, will put us somewhere on a spectrum of, say, depression or ADHD or dyslexia or autism or schizophrenia. Again, this is not saying that being at an extreme of one or more of those spectra cannot be devastating and debilitating. Just that it's not an either/or thing in the way that we currently label people. There aren't some individuals 'on the spectrum' and some not. We are all at different points on the various spectra.

Although there is no reason to doubt Plomin's science, some regard his findings as controversial. When I wrote a review of Plomin's book, I was accused by a commenter of supporting the work of a racist. This is because some believe that Plomin's work could be used to support the ideas of white supremacists (and other racially prejudiced individuals), who suggest that, for example, African Americans tend to earn less and reach lower positions at work because of genetic limitations, not because of social and cultural barriers.

However, if anyone does take such a stance, they totally misunderstand the science. As we discovered in Chapter 2, as far as the genetic side goes, there is far more genetic variation within so-called races than there is between them – this is why the concept of race is so pernicious. As we discovered above, your genetic makeup and environmental development give you the makeup that determines how you will respond to chance and circumstance – but they make no difference to the opportunities you are presented with, which are often driven by cultural and social prejudice and division.

Some even suggest that this science should not be published at all, because it can be misused in this way, but as palaeontologist Henry Gee points out in his book *The Accidental Species*, his work has regularly been misused by creationists to attack evolution because he identifies the gaps in and difficulty of drawing conclusions from the fossil record. It would be ridiculous to suggest that Gee shouldn't describe the reality of palaeontology because his work can be misused by creationists, any more than Plomin should be prevented from presenting the evidence on nature and nurture, just because these facts can be misused by racists.

THE DEPARTMENT FOR CULTURE, MEDIA AND SPORT

The UK government used to have a section called the Department for Culture, Media and Sport (DCMS).* This comes across as something of a ragbag, somewhere to stick the responsibilities that don't quite fit in anywhere else. A department that, dare we say it, doesn't feel quite as important as the big hitters like health or finance or foreign affairs. When I got through my first draft of this book, this section didn't exist – but looking back over the chapter, something seemed to be missing.

I first pinned this down to education and the workplace, but then realised that 'culture' should get a mention too. So, the outcome is a group of missing influences on what makes you *you* that feels like a similar ragbag to the DCMS. However, like those departmental categories, there is a kind of logic in fitting together these other aspects of life that have a significant influence, take up a lot of our time and yet haven't really been covered here.

They are all part of the environment that helps shape you, and help provide life opportunities – but it seems sensible to discuss them separately because of how much of our time they consume, even if, as Plomin suggests, it's hard to pin down the specific influence that education and workplace have on the way that we develop as people. We've seen the cultural aspect coming through already in the discussion of the way we respond to the perception of race, which as we have seen is a cultural rather than a biological distinction.

If you ask a child 'What are you?' they will typically respond with something like 'a girl' or 'a boy' or 'a human being'. But ask an adult 'What are you?' and they will usually tell you their job. 'I'm a plumber', or 'I'm a surgeon'. Even if they've given up the job, they are likely to say 'I'm retired, but I used to be …' In thinking about what makes you what

* At the time of writing it's called the Department for Digital, Culture, Media and Sport. By the time you read this, it will probably be called something else.

you are, it might seem perverse not to have given much attention to your occupation, but this feels like a back-to-front observation.

This is exactly the kind of characteristic that will be partly driven by your genes and partly by a whole mix of things in your environment, including the random opportunities you are presented with. Few of us acknowledge sufficiently how much of that environmental influence is the result of chance and circumstance. We can see the background to being in a particular job, for example, as a forking pathway from the past to the present, where all kinds of decisions and opportunities interact without us having much ability to influence the outcome.

A TREE TO ME

Take my own journey to my role as a science writer. There have indubitably been some genetic influences. I can only think the writing part is genetic because I have always been a voracious reader and an unstoppable writer. I wrote comics as a child and scribbled out my first (thankfully now lost) novel on the train during the commute to and from secondary school. Writing seems to be something I can't avoid doing. Similarly, I suspect the scientific side has a genetic influence – not so much because my dad was a chemist, but because I hung onto the scientific fascination with the world around me and how it worked, a fascination that all younger children seem to have, long after many of my friends at school had gone down other paths.

There were, then, the genetic seeds of 'science writer' all the time. But it certainly was never highlighted in my environmental route through education and career. This started early – at school, on reaching the sixth form, we had to study sciences or humanities. The crossover of science and English simply wasn't a possibility. When I went to university, I thought I was going to do a chemistry degree, but found I enjoyed physics more. As the degree came to an end, I realised that the life of an experimental physicist was not for me, as it seemed to consist mostly of reading data off electronic devices and doing maths, where I preferred something more directly linked to my interests.*

Like many a soon-to-be graduate, I made a desperate raid on the careers office and came away with two possibilities – the scientific side of archaeology, as I'd always been fascinated by stone age sites, or a mathematical discipline called Operational Research (OR), started by physicists during the Second World War and now used to help business and government organisations make better decisions based on data.

Now we get a whole set of potential forks and accidental occurrences. I went to the wrong hotel for an interview for an OR Masters course at Lancaster University. As a result, I was late and had to do the test in the bar. Somehow, the informal surroundings helped me get through it and I got on the course, despite being under-qualified. At the end of my MA, I applied for a number of jobs. One of these I really wasn't interested in – with British Airways. I was stroppy in the interview, which they thought made me more interesting. I got the job. BA was unusual in putting a big emphasis on computing in its OR department, and computing took over my life.

I ended up running the PC department at BA, which meant I regularly spoke to editors of computing magazines. They asked if I'd like to write articles for them, and suddenly writing was coming into my working life – I had always thought of it as a spare-time activity and didn't go looking for it. And best of all for chance as a factor, when writing a book about using the internet in business, I contacted the then-dominant search engine Alta Vista. What I didn't realise was that the American company didn't own the altavista.co.uk web domain – this belonged to a literary agency in London. The agency's owner became my agent and encouraged me to get into popular science writing.

* Plus there was the time I left a soldering iron on in the physics lab over the vacation, nearly burning the place down.

I mention all this to demonstrate how frequently environmental circumstances beyond our control give us a nudge and send us off in a different direction, like a pinball hitting a bumper.

THE TWO CULTURES

We won't spend too much more time on culture and education, except to emphasise again that the wider concept of culture forms a large part of our environment – whether we're talking about the cultural differences between countries or between individuals with different interests. But there is one event in British history that so beautifully encapsulates the impact of culture on our environment that it needs a brief exploration. This is C.P. Snow's 1959 lecture, 'The Two Cultures'.

Charles Percy Snow, a chemist and civil servant, highlighted a cultural divide between the humanities and sciences in the UK. He believed that the establishment, largely drawn from the humanities, tended to disparage the science side. Worse, he accused those from the humanities, who 'by the standards of the traditional culture, are thought highly educated', of simultaneously looking down on the illiteracy of scientists while being themselves almost proud of their scientific ignorance. Snow likened their inability to respond to a question about the Second Law of Thermodynamics as being the scientific equivalent of answering 'Have you read a work of Shakespeare's?' in the negative.

In my experience, certainly, it is still far more common to find a scientist with an interest in the arts than someone from an arts background who shows enthusiasm for the sciences. Although in general the dismissive approach from Snow's time has faded somewhat, it seems possible that the curled lip has remained active where science dares to take a step into the world of the arts. The view seems to be, for example, that science fiction – often written by scientists – is unworthy as literature. Where science fiction does become acknowledged by the literati, it is rapidly labelled as something else. Margaret Atwood, for example, the current favourite SF writer of those with a humanities background, vehemently denies she writes science fiction. In a *BBC Breakfast* interview, Atwood is said to have claimed that science fiction was limited to 'talking squids in outer space'.

The distinction between the humanities and the sciences is surely an important one when we consider what makes us what we are. Some people I speak to who discover I am an author get very interested for a brief moment, only to totally lose that interest when it turns out that I write about science. Others with a science background confess that classical music or literary novels leave them cold.

Of itself, this divide is not surprising – it simply reflects the combination of genetics and environment that makes us who we are. However, this bias that rated the humanities above the sciences seems to have been a cultural split based on 'us versus them' that had no more logical basis than racial prejudice. It was a tribal leftover of a time when science was considered too close to a trade, which the establishment thought beneath it. And though this bias has reduced, it's worth noting, for instance, that backgrounds in the sciences are still dramatically under-represented in the political and media establishment of the present day.

THE MORAL MAZE

One very specific aspect of culture is morality. Like our other personality and behavioural traits, we seem to be predisposed by a mix of nature and nurture in our moral attitude. Our genetic predisposition typically encourages us to favour those with a genetic closeness, though this can be extended to consider the wider community as a kind of extension of our genetic grouping, resulting in altruism. Our cultural and social influences, from a wide range of sources, can be both positive and negative in terms of moral norms.

It can be difficult to pin down how human morality works. At one time it was considered acceptable to experiment on others to try to understand this. A classic example can be found in the experiments run in the 1960s by American psychologist Stanley Milgram. Coming at the time of Nazi war trials, part of the justification for these experiments was to try to understand how large numbers of ordinary people could go along with the Nazi atrocities that occurred during the Second World War – what happened to their 'moral compasses'?

In the experiments, subjects were given the role of 'teachers'. As far as they were concerned, they were taking part in an experiment to understand the effect of punishment on learning. Each teacher was responsible for a 'learner', who was tasked with memorising some text. The learner then had to repeat the text without prompts. When they got something wrong, it was the teacher's role to administer an electric shock, with the aim of seeing whether this shock would make the learner function better or worse.

Unknown to the teachers, the whole experiment was a setup (psychologists have a long history of pretending to be testing for one thing when they're actually testing something else – if you ever take part in a psychology study, assume they're lying to you about the aims). The learner was in on the deceit, and acted out his or her pain. No electric shocks were involved. But the subjects taking the teacher role did not know this.

As the experiment proceeded, the learner got more of the answers wrong. One of the psychology team firmly instructed the teacher to give stronger and stronger electric shocks. The dial controlling this was marked up to a deadly-sounding 450 volts, and the learner became more and more agitated and apparently agonised as the 'voltage' was increased. The general feeling among those in the field before the experiments were run was that only a tiny percentage of subjects would continue as they caused more and more distress and suspected they could cause harm or even death. In reality, 65 per cent of participants went all the way to 450 volts.

The suggestion from Milgram was that the word of authority

tended to overcome moral judgement. If ordered decisively enough to do so, normal individuals would end up travelling down the infamous 'only obeying orders' line to the extent of torturing and killing others. Although there have been some queries as to the validity of Milgram's results, the objections seem to be driven by wishful thinking: the experiments were repeated with considerable success, though with varying percentages, later. This doesn't tell us anything new about where our morality comes from, but does suggest that it is more mutable than some believe.

TROLLEY TRASHING

Thankfully in many ways, though psychologists continue to routinely mislead their subjects, such brutal experiments are now considered unethical, and it's more common to resort to thought experiments, where no one even appears to get hurt in the process. Perhaps the most famous used to explore morality are the trolley problems.

The simplest of them goes like this. You are alone in the control room of a trolley (tram) system and see on the CCTV that an out-of-control trolley is heading for a group of five people standing on the line. There is no way to warn the people, or for them to move off the track in time to escape. However, between them and the trolley is a switch (set of points), which can be controlled by you. With the touch of a button, you can divert the trolley onto another track and save the lives of those five people. Unfortunately, there is one individual standing on the second track. This means that if you divert the trolley, that person, who was previously safe, will die.

The question you are asked to consider is whether you would leave things to play out, meaning that the five people would die, or you would press the button, saving the five people but causing the death of another individual. Please try to determine what you would do before moving on. There is no right or wrong answer.

Before we consider your response, let's try another trolley problem. Once again, a runaway trolley* is careering down the track towards five people. Once again, you don't have time to stop it or warn them. But now things aren't so simple. You are on a bridge, just before the switch in the track. There isn't time to get to the nearest control for the switch, which is beside it on the ground, to change the trolley's direction. Even if you threw yourself off the bridge, you wouldn't be heavy enough to trip the heavy-duty control lever. However, there is an extremely heavy person, foolishly perched on the parapet of the bridge. If he were pushed off the bridge, he would land on the control and his weight would be sufficient to move the switch, saving the five people. Unfortunately, the fall would kill him. Would you push him to his death, or leave the five people to die? Again, take a moment to decide before reading on.

Bearing in mind there genuinely is no right or wrong answer, the majority of people would throw the switch in the first case, but would not push the person off the bridge in the second example. Yet morally, the circumstances are identical. In each case, if you take action the result is that five people who would otherwise die are saved, at the price of killing someone who would otherwise have lived.

The psychologists behind these tests suggest that your morality is conditioned by how direct your involvement appears to be with the action. This, it's suggested, is why it seems to be easier to kill someone with a gun, and even more so at long range with a rifle, than close up and personal with a knife. In the first trolley experiment, you are dispassionately pressing a button, which results in a distant action; in the second case you are actively pushing someone to his death. Yet the fact that these actions indubitably have identical consequences demonstrates that your moral compass is not set in stone, but free to be influenced by circumstance. If you are remote, your decisions are more influenced

---

* Someone really ought to sue this company.

by rational thinking; when you are up close and personal, emotion becomes a more significant driver.

I suspect the psychologists are correct, but there is a danger of reading too much into the specific thought experiment just because it is that – a thought experiment, where it is very difficult to be certain of the causality behind the decision. In the second case, the circumstance is far more contrived than the first, and it is impossible to see how it could work in reality. How could you possibly know that the heavy person would successfully trip the control? For that matter, how could you know that the person would land on exactly the right spot?* You couldn't, and that makes it far easier to reject this option, as you could well end up killing the one person without saving the others.

In reality, just as the environmental contribution to your personality is not clear-cut, but pulled together from a complex interaction of small factors, so real-life moral decisions are neither clearly logical nor purely emotional. Your moral decisions are far more likely to be the result of many, sometimes conflicting, considerations.

CAN WE EVEN DEFINE 'YOU'?

We are likely to always be limited, then, in finding definitive causes for your particular personality and behaviour. However, some suggest that there really isn't anything as fixed as we might imagine waiting to be uncovered – that, in effect, there isn't a fixed 'you' beneath the surface. In his book *The Hidden Half*, Michael Blastland recounts the 'provocative ideas' of Nick Chater, Professor of Behavioural Science at Warwick Business School.

Chater suggests that rather than settling in adulthood to a particular character and collection of opinions that together form the deep-rooted

* In the original version of the problem the heavy person is so heavy that when you push them onto the track and the trolley runs into them, they bring the trolley to a stop. This is just bad physics.

basis of what you are, instead you are pretty much entirely on the surface with no deep roots at all. This theory suggests that what is usually presented as the bedrock nature of a person's identity is, in fact, far more fluid than most of us (including most academics) believe. Chater tells us that rather than being a settled and identifiable thing, the internal *you* is constantly fluid, switching and changing as is necessary to respond to circumstances.

As evidence of this occurring, Blastland gives an example of a demonstration he undertook for a radio broadcast, replicating earlier experiments in Sweden, where participants were asked about their political views on a scale of 1 to 10. In a break in proceedings, the programme's organisers switched round the participants' answers so that anything that they scored in the 3 to 7 range was swapped from negative to positive and vice versa. The participants were then given the doctored answer sheets and asked to defend their views. Around three-quarters of them happily defended the opposite viewpoint to the one they had actually put down.

The suggested inference is that this result is supportive of Chater's thesis; however, there is an issue. Limiting the experiment to answers that scored in the 3 to 7 range inherently selected shallowly held views. In essence, the experiment shows that we are only slightly attracted to views which we only lightly support, perhaps because we are unsure on these topics. Is this really a surprise? Blastland goes on to comment:

> Even if we're certain that we prefer beer to spirits, one political party to another, tax cuts to more money for state healthcare and it will take an earthquake to change our minds – certainly not some piece of surface-level trivia – such 'deep' preferences are evidently not a reliable characterization of how the mind works.

However, to use one of those examples, as someone who enjoys beer and can't stand spirits, I would never simply switch the argument because

I thought I'd written down the opposite. If I had been answering such a question, I would have given 10 to beer – my certainty would have excluded my answer from manipulation. The experiment tells us nothing about deep preferences where it would 'take an earthquake to change our minds', only about vague partialities. As much more of a floating voter politically than I am on the choice between beer and spirits, I would be more likely to give a political party a middling score, and would then have been able to justify voting for one of several parties (as I have in the past).

It's not that the interpretation of Blastland's experimental results, or Chater's idea, is entirely incorrect, and the experiment does emphasise just how fluid we are in our opinions on topics in which we aren't strongly invested either way. If this weren't the case, elections would be very boring, because everyone would always vote for the same party that they had always supported and the outcome would only vary as the voting population changed. And it's true that the way that the participants in the experiment were prepared to shift viewpoint over such a short period of time was particularly impressive.

What, perhaps, we can conclude is that there are aspects of opinion and character where you are likely to be absolutely solid and others where you will be more fluid than you typically admit. The 'you' that is presented, for example, to your family, is likely to be quite different to the 'you' that your co-workers experience. That's a variation over context, but similarly as these experiments show, there can be considerable variability over short periods of time too. This also reflects that, perhaps thankfully, many of us don't see every issue in black or white but that we can appreciate both sides of an argument.

At the time of writing, the UK is tearing itself apart over the decision to leave the European Union. Parliament itself is unable to make any sensible decisions, merely acting to avoid them. Vast swathes of the country are either for remaining in the EU or leaving – many of them strongly polarised. They would not have rated the options in the 3 to 7 range in Blastland's questionnaire. Both sides of the divide ignore evidence unless it supports their viewpoint. At the time of going to press, the departure has been finalised and politicians have expressed hope that the country can be 'brought together', but there remains a significant split in the country and difficult political times ahead. I suspect that on issues like this, it's not that we are more fluid in our opinions than we think, but rather we are far too rigid to achieve sensible outcomes.

Many of the decisions we face in our complex world do not have single 'right' answers. When I have run creativity seminars for business people, something I have always emphasised is that the search for *the* definitive right answer is a fool's errand. Most real-world situations have many right answers – and just as you may fancy five different things on a restaurant menu, you might equally feel that apparently opposing viewpoints both have positive aspects. We are always choosing between imperfect selections with imperfect knowledge and it's no surprise that, despite retaining some deeply-held beliefs and ideas, the whole that is 'you' also contains many contradictions and fluid possibilities. It's just a shame that this flexibility doesn't extend to some of the important political decisions we face.

So, with these final factors we are heading to the final synthesis – pulling together every little thing that makes you *you*.

10

EVERY LITTLE THING

We have seen how many different components and pathways into the past have made 'you' possible and all of these are necessary contributors to your being. As a human, you are the most complex entity in the known universe. I know biologists don't like us to consider humans exceptional – but you truly are.

THE MANY STRANDS THAT LEAD TO YOU

Although we opened by showing how limited the concept of genealogy is in getting a good picture of what you are and where you've come from, it's worth remembering that in the end, genealogy was a precursor to genetics – and so it was a way that gave a first inkling of human genetic composition.* It's just that looking back and seeing who you were related to hundreds of years ago gives a narrow view, in part because of the dilution of impact across all those generations, and in part because of the exponential explosion. But, as nature versus nurture experiments have shown, we certainly can't ignore the hugely important genetic aspect of what makes you who you are.

* Both 'gene' and 'genealogy' have the same Greek root.

We've also looked at the atoms that make up your body, the energy sources that power it, the origins of life and human beings, the environment (including culture, education and the workplace) and the technological surroundings that help shape you and, perhaps the most fundamental thing that is 'you', that fuzzy, flimsy, hard-to-pin-down thing that is your consciousness.

Take a moment to assess what it feels like to be 'you'. Just by considering what 'you' is, it's inevitable that your consciousness gets engaged – but what else do you think of? It's a complex mix that is likely to encompass many of the topics we've already covered, but there are likely to be others as well.

WHAT'S MISSING?

For example, we have not covered your hobbies and activities, yet for some these can be deep-rooted parts of what makes them who they are. Whether it's following a football team, researching local history, looking after a pet, jogging or bicycling, music ... it can feel that a personal interest is an essential part of you. I am always fascinated when people who are asked about spare-time interests say, for example, 'I love my music.' I enjoy music, particularly singing and listening to Tudor and Elizabethan church music,* but I don't feel such a deep attachment that makes what I listen to 'my music'. I like music, but could live without it, while some would describe their spare-time activities as *being* their life.

We're not the only animals that indulge in recreational activities, over and above the biological essentials of food, drink, sleep, socialising and procreation, or the human extension of these in creativity that often becomes our work. Other animals do things for the fun of it – you only have to watch a video of otters playing in the snow to see this. But we have to be more than a little careful about ascribing a human inner voice

* Demonstrating that I am down with the kids.

to other species, however cute the behaviour may seem to us. Remember the problem of understanding what it is like to be a bat.

I have recently seen an internet video of a bird 'playing' with a golf ball. The social media description ran 'This bird just discovered that golf balls bounce on concrete and it's the cutest thing'. We see a large bird picking up a golf ball, running over to a path and hurling the ball to the ground. When it bounces, the bird jumps back in apparent amazement, then keeps going back for another go. It's easy to interpret this as the bird enjoying the experience of bouncing the ball, because that's what a human would be doing in these circumstances. In reality, of course, the bird is trying to smash what it probably assumes is an egg to get to food. It's not having fun, it's probably getting extremely frustrated by the ball's failure to crack.

Although we haven't covered hobbies explicitly, like culture they fit in as part of the environment that makes you what you are, driven by the personality traits we've already covered – and have some similarities with our consideration of education and the workplace. The same goes for another apparent omission – having children.

Some friends of ours have just had their first baby. He's lovely – and he's absolutely the central, driving factor of their lives right now, just as I remember being the case when our own children arrived. Up to this point, both members of the couple had got a lot out of their careers and their spare-time activities, but for the moment, this small, new life has taken over everything, from their sleep pattern to their daily routines. That will change, of course, as family life becomes normalised – but children will alter their lives for ever. It's not surprising that our children feature large in what many of us think of as our defining aspects.

However, it didn't seem necessary to give children a separate chapter, because having children is a basic biological imperative. This is not a criticism if, for whatever reason, you do not have children – that's now part of what makes you *you*. Nor is it saying that everyone wants to be a parent or should want to do so. It is just that for those who do have children, there is a strong biological aspect to this. The relationship parents have with their children is tied so powerfully into the genetic aspects we've already covered that it doesn't need a separate consideration. I'm not downplaying the importance of family. Rather, I'm saying that though the impact of having children on their parents' lives is deep and long-lasting, it is not necessary to cover it as a separate factor.

If there's something else that I haven't covered, that's fine. We all have huge overlapping similarities – but equally, each of us is different from the rest. We have seen how a combination of lots of tiny environmental differences can result in very big differences in outcome. There will be other factors that feel, perhaps, more important to you than they do to me. But I hope that in joining me on this voyage of discovery into what makes you *you*, you now feel that you have a better understanding of the complex, fascinating, wonderful thing that is the particular human being who is reading these words.

Whatever you think that you are, whatever it is that makes you that way, you are certainly truly remarkable

FURTHER READING

Inevitably, a book like this can only give tasters of aspects of the science behind you being *you*. If you would like to investigate in more depth, here are some books that explore the areas covered in the individual chapters in more detail, with additional sources to zoom in and gain more depth on specific points.

YOUR ANCESTORS WERE ROYAL

High level

A Brief History of Everyone Who Ever Lived, Adam Rutherford (Weidenfeld & Nicolson, 2016) – A wide-ranging exploration of the genetic contribution to our individual makeup and family tree from an excellent storyteller.

Superior: The Return of Race Science, Angela Saini (Fourth Estate, 2019) – shows how science has been misused to construct the fiction of racial inequality.

Zoom in

How many people have ever lived? – the figure of 108 billion is based on the 2017 PRB estimate, detailed here: www.prb.org/howmanypeoplehaveeverlivedonearth/

Upper, middle and lower class in the UK: 'Class System', 1966 sketch from satirical TV show *That Was The Week That Was* featuring John Cleese, Ronnie Corbett and Ronnie Barker www.youtube.com/watch?v=9tXBC-71aZs

Genetic study showing similar results to Chang's on time to European common ancestors: Peter Ralph and Graham Coop (2013). 'The

Geography of Recent Genetic Ancestry across Europe', *PLoS Biology*, 11(5), p.e1001555.

More sophisticated modelling for common ancestors using population constraints: Douglas Rohde, Steve Olson and Joseph Chang (2004). 'Modelling the recent common ancestry of all living humans', *Nature*, 431(7008), pp. 562–566.

Original paper using simple models to estimate time to common ancestors: Joseph Chang (1999). 'Recent common ancestors of all present-day individuals', *Advances in Applied Probability*, 31(4), pp. 1002–1026.

Study showing the genetic variation within populations is much greater than between populations: Noah Rosenberg et al (2002). 'Genetic Structure of Human Populations', *Science*, 298(5602), pp. 2381–2385.

STARDUST MEMORIES

High level

30-Second Elements, Eric Scerri (ed.) (Icon Books, 2013) – a good, simple, high-level introduction to the chemical elements.

The Disappearing Spoon, Sam Kean (Black Swan, 2011) – doesn't cover all the elements, but great storytelling in a meander through some of them.

Before the Big Bang, Brian Clegg (St Martin's Press, 2009) – good exploration of theories of the origin of the universe and where matter came from.

Zoom in

Constituents of human body: H.H. Mitchell et al. (1945). 'The Chemical Composition of the Adult Human Body and its bearing on the biochemistry of growth', *Journal of Biological Chemistry*, 158(3), pp. 625–637.

Silicon equivalent of benzene: Kai Abersfelder, Andrew White, Henry Rzepa and David Scheschkewitz (2010). 'A Tricyclic Aromatic Isomer of Hexasilabenzene', *Science*, 327(5965): 564–566.

Value of elements in human body (high): bgoodscience.wordpress.com/2011/03/21/body-for-sale-how-are-your-chemical-components-worth/

Value of elements in human body (low): www.thoughtco.com/worth-of-your-elements-3976054

Value of elements in human body (typical): www.datagenetics.com/blog/april12011/

WHERE DID THE SPARK COME FROM?

High level

The Immortal Life of Henrietta Lacks, Rebecca Skloot (Macmillan, 2010) – fascinating account of the life, and cellular afterlife, of a woman who changed the medical world.

The Vital Question, Nick Lane (Profile Books, 2015) – examines the detailed possibilities for the origins of life and the move from simple to complex life forms.

Zoom in

Early atmosphere deduced from zircons: D. Trail, E.B. Watson, N.D. Tailby (2011). 'The oxidation state of Hadean magmas and implications for early Earth's atmosphere', *Nature*, 480(7375): 79–82.

Paley's watchmaker argument: *Natural Theology*, William Paley (Cambridge University Press, 2009) – often quoted but rarely read.

Life, entropy and energy: Jeremy England (2013). 'Statistical physics of self-replication', *The Journal of Chemical Physics*, 139(12), p. 121923.

YOU ARE WHAT YOU EAT

High level

30-Second Energy, Brian Clegg (ed.) (Ivy Press, 2018) – straightforward illustrated introduction to energy.

There is no Planet B, Mike Berners-Lee (Cambridge University Press, 2019) – an environmental handbook, of particular interest in the way it analyses food energy chains.

SOS, Seth Wynes (Penguin Books, 2019) – overly simplistic, but useful in its presentation of levels of carbon emissions from various activities.

Zoom in

Water use and agriculture: 'Managing water sustainably is key to the future of food and agriculture', OECD, www.oecd.org/agriculture/topics/water-and-agriculture/

Global temperature changes: 'Global Temperature', NASA, climate.nasa.gov/vital-signs/global-temperature/

Sea level rise projections: 'Sea Level Change', IPCC, www.ipcc.ch/site/assets/uploads/2018/02/WG1AR5_Chapter13_FINAL.pdf

A DIFFERENT APE

High level

The Accidental Species, Henry Gee (University of Chicago Press, 2013) – a great exploration of what the fossil record can (and can't) tell us about human evolution.

30-Second Evolution, Mark Fellowes and Nicholas Battey (eds) (Ivy Press, 2015) – a simple, illustrated introduction to evolution and how it works.

Zoom in

Space impactors: *Cosmic Impact*, Andrew May (Icon Books, 2019) – excellent introduction to asteroid, comet and meteor impacts on the Earth.

Out of Africa mitochondrial DNA: Rebecca Cann et al. (1987). 'Mitochondrial DNA and human evolution', *Nature* 325(6099): 31–36.

Newspaper story about MRD being our 'oldest ancestor': *i* newspaper, 26 August 2019.

The significance of sea and freshwater food in the development of the hominin brain: *Human Brain Evolution*, Stephen Cunnane, Kathlyn Stewart (John Wiley, 2010).

IS YOUR CONSCIOUSNESS AN ILLUSION?

High level

Consciousness: A Very Short Introduction, Susan Blackmore (Oxford University Press, 2017) – a straightforward and approachable introduction to the concept.

The Quantum Age, Brian Clegg (Icon Books, 2014) – a readable introduction to quantum physics and how we make use of its strange behaviour in our technology.

Consciousness: An Introduction, Susan Blackmore, Emily T. Troscianko (Routledge, 2018) – a more heavyweight coverage of the nature of consciousness.

Incognito: the secret lives of the brain, David Eagleman (Canongate, 2011) – an exploration of the functioning of the brain that goes beyond consciousness.

Zoom in

Bats: Thomas Nagel (1974). 'What is it like to be a bat?', *The Philosophical Review* 83(4): 435–450.

Consciousness delay: Benjamin Libet, *Mind Time* (Harvard University Press, 2004) – Libet's own book is technical but gives the insider view on his work.

Alternative explanation for apparent conscious delay: Aaron Schurger et al. (2012). 'An accumulator model for spontaneous neural activity prior to self-initiated movement', *Proceedings of the National Academy of Sciences*, 109 (42): E2904–13.

Intuition in decision-making: Antoine Bechara, Hanna Damasio, Daniel Tranel and Antonio Damasio (1997). 'Deciding advantageously before knowing the advantageous strategy', *Science* 275(5304): 1293–5.

Free will in quantum and chaotic systems: Brian Clegg, *Dice World* (Icon Books, 2013) – an approachable guide to randomness and probability and their impact on our lives.

Choosing cards: Cosmides and Tooby, 'Cognitive Adaptations for Social Exchange'. In Barkow, Cosmides and Tooby (eds), *The Adapted Mind* (OUP, 1992), pp 163–228 – detailed exploration of the social impact on brain function.

Quantum teleportation: Brian Clegg, *The God Effect* (St Martin's Press, 2009) – an introduction to quantum entanglement and its various implications, including quantum teleportation.

Conscious machines: John Pavlus: 'Curious about consciousness? Ask the self-aware machines', *Quanta Magazine*, July 2019, accessed online at https://www.quantamagazine.org/hod-lipson-is-building-self-aware-robots-20190711

Resonance theory of consciousness: Tam Hunt, 'Could consciousness all come down to the way things vibrate?', *Interalia Magazine*, November 2018, accessed online at https://www.interaliamag.org/articles/could-consciousness-all-come-down-to-the-way-things-vibrate

LIFE IS MORE THAN BIOLOGY

High level

Upgrade Me, Brian Clegg (St Martin's Press, 2008) – looks at the way we have enhanced ourselves to make ourselves more than human, from the simplest technology to brain implants.

Factfulness, Hans Rosling (Sceptre, 2018) – a fascinating exploration of how and why we think the world is much worse than it really is.

The Perils of Perception, Bobby Duffy (Atlantic Books, 2018) – like Rosling, Duffy looks at the gap between perception and reality, but with less focus on our attitude to the 'developing vs developed' world.

Zoom in

The failings of GDP: 'Unattributed: The trouble with GDP', *The Economist*, April 2016, accessed online at www.economist.com/briefing/2016/04/30/the-trouble-with-gdp

Incorrect initial data on EU immigration to the UK: Office for National Statistics, *Understanding different migration data sources: August 2019 progress report*, www.ons.gov.uk/peoplepopulationandcommunity/populationandmigration/internationalmigration/articles/understandingdifferentmigrationdatasources/augustprogressreport

Bacon on health (and much more): *Letter Concerning the Marvellous Power of Art and Nature and Concerning the Nullity of Magic*, Roger Bacon, trans. Tenney L. Davis (Kessinger Publishing, 1940).

THEY DON'T MESS YOU UP, YOUR MUM AND DAD

High level

Junk DNA, Nessa Carey (Icon Books, 2015) – fills in details of the role of DNA outside of the genes and how it controls genes and more.

Blueprint, Robert Plomin (Allen Lane, 2018) – detailed but approachable description of Plomin's working showing the influence of nature and nurture on our personalities.

The Hidden Half, Michael Blastland (Atlantic Books, 2019) – fascinating exploration of the half of the world that we can't easily pin down and explain.

Zoom in

Influence of genes on the microbiome: Julia Goodrich et al. (2017). 'The relationship between the human genome and microbiome comes into view', *Annual Review of Genetics*, 51: 413–433.

Variation in 'identical' crayfish: Günter Vogt et al. (2008). 'Production of different phenotypes from the same genotype in the same environment by developmental variation', *Journal of Experimental Biology* 211(4): 510–23.

Long-term study on boys from deprived backgrounds: *Shared Beginnings, Divergent Lives*, John Laub and Robert Sampson (Harvard University Press, 2003).

Electric shock experiment: Stanley Milgram (1965). 'Some Conditions of Obedience and Disobedience to Authority', *Human Relations* 18(1): 57–76.

PICTURE CREDITS

Page 9: based on https://www.townandcountrymag.com/society/a20736482/british-royal-family-tree/; page 31: International Atomic Energy Agency; page 54: Creative Commons image by YassineMrabet; page 63 (top): Creative Commons image by Michael/Jacopo Werther; page 79: Creative Commons image by A Loose Necktie; page 91 (both images): Creative Commons images by Chiswick Chap; page 97–9: based on illustrations in *The Accidental Species* by Henry Gee (reproduced with permission); page 110: Alamy; page 124: image by Edward H. Adelson; page 129: Tyler Vigen (www.tylervigen.com); page 138 (upper) Creative Commons image by Arkopri; (lower) Creative Commons image by Sakurambo; page 166: Creative Commons image by Stephencdickson.

INDEX